CAMBRIDGE COUNTY GEOGRAPHIES

General Editor: F. H. H. GUILLEMARD, M.A., M.D.

KENT

Cambridge County Geographies

KENT

by

GEORGE F. BOSWORTH, F.R.G.S.

With Maps, Diagrams and Illustrations

Cambridge:
at the University Press
1909

CAMBRIDGE UNIVERSITY PRESS
Cambridge, New York, Melbourne, Madrid, Cape Town,
Singapore, São Paulo, Delhi, Mexico City

Cambridge University Press
The Edinburgh Building, Cambridge CB2 8RU, UK

Published in the United States of America by Cambridge University Press, New York

www.cambridge.org
Information on this title: www.cambridge.org/9781107660045

First published 1909
First paperback edition 2013

A catalogue record for this publication is available from the British Library

ISBN 978-1-107-66004-5 Paperback

CONTENTS

ILLUSTRATIONS

ILLUSTRATIONS

The Illustrations on pp. 4, 7, 32, 34, 40, 49, 55, 65, 70, 72, 81, 100, 108, and 118 are from photographs by Messrs F. Frith & Co., Ltd., Reigate; and those on pp. 59, 96, 105, 113, and 125 are from photographs by The Homeland Association, Ltd.

1. County and Shire. The Word *Kent*. Its Origin and Meaning.

It has been well said that our national history is made up of local history, and that our knowledge of the history of England as a whole will be all the better if we learn something of the way in which the English kingdoms were formed. This will help us to understand the relation which our modern divisions bear to the ancient ones. These modern divisions are named counties and shires, and we call one Kent and another Stafford-shire. In the latter instance, we note the affix *shire*, while in the former there is not this special ending. Let us endeavour to find out the reason for this difference, and we shall then be in a better position to understand the origin of the county of Kent in the early days of our history.

Look carefully at a map of England and make a list of the divisions that end in *shire*. It may at once be said that these are portions or *shares* of a larger division. Thus Staffordshire was once a part of Mercia, one of the great kingdoms in early English days. Again Berkshire and Gloucestershire were formerly parts of Wessex, another English kingdom. Now look at the map and pick out the divisions that do not end in *shire*. Of these it may generally be said that they are the survivals of the old English kingdoms, which have kept their former extent and in some cases their original names.

Perhaps we could not take two better counties than Sussex and Kent to illustrate this fact. Both these counties were originally kingdoms and have retained their boundaries and names from the earliest times when the Saxons and Jutes came to settle in England.

The history of England tells us that our English forefathers divided our land into several kingdoms, of which Kent and Sussex were two ; so that, for fourteen hundred years, these two counties have kept the names that they now bear. That is a very remarkable fact, and one of the deepest interest for us who are going to read about the geography of Kent. History and geography have a very close connexion at times, and here the one subject helps to illustrate the other.

The very word Kent has a history that carries us back to a period before the invasion of Julius Caesar. While most of our present English counties have English names, Kent stands almost alone in bearing one of Keltic origin. This fact bears witness to its antiquity, and leads us to understand that there is much in a name. Norfolk, Suffolk, Essex, Sussex and Middlesex are all good English words, whose meaning is evident at a glance. But with Kent the case is entirely different, and one has to learn a good deal of history to know how it got its name and why it has kept its name.

Pytheas, who lived about 350 B.C., was one of the earliest explorers who visited our land, and he mentions Cantion as one of the places he visited. Ptolemy, who flourished about 150 A.D., and was one of the greatest of ancient geographers speaks of Cantium, which may be said, roughly, to be represented by the modern Kent. In those early times we may safely say that the Kelts were living in England, and so it comes about that Kent is derived from *Caint*, a Keltic word meaning the open country, and was given to the long slip of land lying along the sea-shore and the Thames.

In the *English Chronicle*, Caint becomes *Cantwara land* and *Cent*, and in the *Domesday Book* it is written *Chenth*. In later histories it takes the form Kent, as you see it on the map of England at the present time. There is one other fact of interest that may be mentioned. Kent has two cathedral cities—Canterbury and Rochester—and this probably arose because, in early English times, it was subdivided into two kingdoms—East and West Kent. Canterbury was the capital of East Kent, and its name Cant-wara-byrig means "the town of the men of Kent." The Archbishop of Canterbury signs his name *Cantuar*, which is simply a contraction of Cantuariensis the Latinised title of the See.

Canterbury Cathedral

2. General Characteristics. Its Position and Natural Conditions.

It is generally admitted that Kent is one of the most interesting of our English counties, and there are many reasons why this county should be the first to be noticed in studying the geography of England.

Kent has been the scene of some of the most note-worthy events in our history, about which we shall read in later chapters. Here it may be mentioned that it was the first English landing-place of Julius Caesar and his Roman army ; Hengest and Horsa, the leaders of the Jutes, first conquered Kent and settled in it ; and, at a later date, Augustine first set foot in Kent to Christianise it.

Again, Kent is the land first seen by the majority of visitors from the Continent, and they travel through the " Royal " county to reach the metropolis. Kent may be regarded as the corner-stone of the kingdom, guarding it with chalk cliffs, " the white walls of old England."

The broad estuary of the Thames, washing its northern shore, is covered with vessels that carry to London the riches of all parts of the world, while Kent itself has numerous ports that have a large and increasing trade. Thus we see that Kent has the first claim on our attention, because of its importance in our national history, and also owing to its favourable situation for trade and commerce.

Kent is a maritime county in the south-east of England. It is nearer to France than any other portion of England,

and the Strait of Dover is only about 20 miles across at its narrowest part. The extent of the Kentish coast-line is considerable, and measuring both the open sea-board and the estuary of the Thames and Medway, there is a total length of water-line of about 140 miles.

Kent is one of our agricultural counties, and has always been famous for its corn and hops, apples and cherries, sheep and deer. It has been well named the "Garden" county, and a French visitor to England, more than 200 years ago, remarked that the grass seemed to be finer and of a better colour than elsewhere. After a long and glowing account of Kent, he thus concludes, "The eye cannot but be much delighted with the natural and even neglected beauty of the country, and the English have reason to value it."

In every way Kent has great advantages as an agricultural county. Its soil is fertile; its climate is equable; and there are special facilities, both by land and water, for carrying its produce to suitable markets.

The north-western portion of Kent is now part of the county of London, and in this small area is a large industrial population, who either work in the metropolis or gain their livelihood in the boroughs of Deptford, Greenwich, Woolwich, or Lewisham.

We know how the position of Great Britain as the centre of the land hemisphere has affected its history, and made it the greatest maritime nation. In a similar way, we may attribute the wealth and prosperity of Kent to its fine position, its nearness to the Continent, its length of coast, and its fertility of soil. Kent may be considered as one of the windows through which England looks into the great world.

The scenery of Kent is very varied, and shows a pleasing succession of hill and dale. There are many interesting points of view in the county, and fine prospects will charm the spectator who stands on Dover cliffs, on Shooter's Hill, near London, or on the hill behind the little church of St Martin's, Canterbury.

Shakespeare Cliff, Dover

3. Size. Shape. Boundaries. Detached Portion.

Kent, which occupies the south-eastern extremity of England, may be considered a peninsula with two promontories. The northern promontory is formed by Foreness and the North Foreland, while the southern promontory is Dungeness.

The length of the county is 64 miles, if measured from London to the North Foreland, and the breadth, measured from the North Foreland to Dungeness, is 38 miles. The circumference of the county is about 170 miles, and this encloses an area of 995,014 acres or 1554 sq. miles.

In point of size, Kent is the ninth English county, and embraces an area about one thirty-third of the whole of England. It is interesting to note that Kent is a little larger than Essex, rather smaller than Somerset, and more than twice as large as any other county in the Thames basin.

If we compare the shape of Kent with that of other English counties, we shall find that it is more compact, and is roughly the shape of a quadrilateral.

The boundaries of Kent on the north are the estuary of the Thames and the North Sea, while on the east and south-east the county is bounded by the North Sea, Strait of Dover, and the English Channel. On the west, Kent is bordered by Surrey, and on the south-west by Sussex and the river Rother. It will thus be seen that on the

west and south-west the boundaries are mainly artificial,
while on the other sides they are natural.

There is one peculiarity connected with Kent that
may be noticed in this chapter. A part of Kent is
situated in Essex, and is known as North Woolwich.
There are other counties where the same thing occurs,
but it is very difficult to get a satisfactory explanation of
this fact. Why a portion of Kent should be in Essex
has given rise to many theories, none of which is entirely
acceptable. Perhaps this isolated portion of Kent in Essex
dates from the time when Essex and Kent formed one
kingdom. When the separation came about, it was
probably arranged, for some reason or other, that Kent
should resume its former territory plus this district in
Essex. One of our historians gives what is probably the
best reason for part of Kent being in Essex. He says
that Count Haimo, Sheriff of Kent in William the Conqueror's reign, had land on both sides of the river at
Woolwich, and in this way the property on the north
bank in Essex became included in the county of Kent.

There are also some parishes that have outlying
portions in other parishes. It is given as an explanation
that as some of these isolated spots are in the centres of
forests, or in places where there were forests in ancient
times, the inhabitants of the lowland, or open land, had
parts of forests given them for purposes of fuel ; and that
when the trees were cut down, and the place left bare
in the middle of the forest, they claimed that land as
part of their own parish.

It has now been settled, by Act of Parliament, that

these outlying portions may be joined with the district in which they are situated, if both parties interested in the locality are agreed to the amalgamation.

4. Surface and General Features.

We shall understand the succeeding chapters in the geography of Kent all the better if we first get a good

The Weald of Kent

idea of the chief features of its surface. The surface of Kent is quite English in character, for, although there are no mountains or hills of any height, there is considerable variety in its physical features. There are marshlands along the Thames and also in the south;

there are breezy downs and fertile valleys; broad plains and the region known as the Weald, which was once part of an extensive forest.

The most interesting feature of the county is the range of hills known as the North Downs. They enter the county from Surrey, a little north of Westerham, and extend to the chalk cliffs of Dover, where they terminate in Shakespeare Cliff, a bold and picturesque headland made famous by Shakespeare's reference to it in *King Lear*. The North Downs may be called the backbone of Kent, although not a continuous range. A reference to the map will show that it is broken in three places in Kent, by the rivers Darent, Medway, and Stour.

The North Downs occupy the main portion of the county and contain the highest points. The chief elevations are within a short distance of Folkestone, where Paddlesworth Hill and some others attain a height of over 600 feet; and near Westerham, where the Downs reach their maximum elevation in Betsom's Hill, 811 feet high.

South of the Thames valley there is a district with some wooded hills. Greenwich and Blackheath are familiar names, and Shooter's Hill near Woolwich is 424 feet in height. This, however, is a region of little importance when compared with the North Downs as a whole.

The North Downs really divide Kent into two parts. To the north of the Downs runs a narrow strip of lowland, broken by river estuaries and marshy islands, such as the Isle of Sheppey, and, for the most part, devoid of interesting scenery. Between Dartford and Gravesend many manufactures are carried on. To the south of the Downs lies a belt of undulating country belonging to the Wealden district. This lower land is separated from Sussex by the river Rother, and ends at the English Channel in Romney Marsh.

The Weald (*weald* = forest) is a large tract, extending from Romney Marsh to Surrey. In ancient times it was a forest, known as Andred's-Wald. Most of the forest land has been cleared, and towns and villages have sprung up. There are still, however, some beautiful woodlands, and the whole district is very picturesque. Indeed the Weald is called the "Garden" of Kent, and well it deserves this name. There are rich cornfields, smiling pastures, fruitful orchards, and beautiful hop-gardens; and the comfortable homesteads, with substantial farm-buildings, hop-oasts and well-stored stack-yards, all go to show that Kent is one of the most fruitful of our counties.

The district known as Romney Marsh has an area of about 45,000 acres. It extends from Hythe to Rye, a distance of 18 miles, and has a breadth of about 12 miles. Romney Marsh also includes Walling, Denge, and some other marshlands, and embraces twenty parishes. The Marsh is protected from the sea by Dymchurch Wall, and, but for this, the sea would overflow it at once. The drainage of the Marsh is effected by a number of divisions called "waterlings," and the district is now provided with good roads. Cattle and sheep are reared in great numbers, for there is abundance of good pasture. The sheep are a peculiar breed, hardy, and able to endure privations from the cold and damp.

There is no doubt that the whole of Romney Marsh was once covered by the sea. Perhaps it was first reclaimed by the Romans ; and when the Saxons settled here, they were called the "Marsh" men. They, however, did little to protect the marshes from the inroads of the sea ; but there are records of this work being undertaken at a later date by the Archbishop of Canterbury, who owned this land.

Dymchurch Church

In a later chapter we shall have something more to say about the building of the sea-wall, and its repair by the "Lords of the Levels." Here we would remark that although Romney Marsh is not, at first sight, one of the most beautiful parts of Kent, yet it has features

that appeal to the best affections of our nature. The
wide stretches of rich grass-lands, the flocks of docile
sheep, the hospitable dwellers in this out-of-the-way
peninsula, and the number of its churches all go to
show that Romney Marsh has a character of its own,
both in the physical and economic life of Kent.

5. Watershed. Rivers.

From the previous chapter it will be gathered that
the watershed of Kent is the chalk range of the North
Downs. This line of hills is cleft by several rivers,
flowing northward to the Thames. From west to east
the Thames receives the following feeders from Kent :—
Ravensbourne, Cray, Darent, and Medway. Of course,
the Thames cannot be called a Kentish river, but as it
bounds the northern shore of the county it will be
specially noticed in the chapter on the coast of Kent.
Here we need only note that the estuary of the Thames
extends to the Nore Lightship; and that from London
to the Nore "the Thames is the world's exchange."
There is a ceaseless passage of vessels to and from all
parts of the world ; and, although this part of the
Thames is not beautiful as regards its scenery, its commercial importance is unrivalled.

The most important of all the Kentish rivers is the
Medway. It is the finest tributary of the Thames and
has a total length of 60 miles. A glance at the map
will show that it is the great outlet for Mid-Kent.

Upnor Castle

From its mouth to a spot between Aylesford and
Maidstone the Medway is tidal, although formerly the
tide flowed a mile or two above Maidstone. The
Medway has four sources, two of them being in Sussex,
one in Kent, and one in Surrey. One of these streams
rises near East Grinstead in Sussex, and flowing to Penshurst, it receives the Eden, a little mill-stream. It then
proceeds by Tonbridge to Yalding, where it is increased
by the waters of the Teiss and Beult. The main stream
now pursues its course past Maidstone to Rochester, where
the estuary is very picturesque. For the last twelve miles
of its course the estuary of the Medway is deep and nearly
a mile across. On its way to Sheerness it passes Chatham
and Upnor Castle, and enters the Thames between the
Isles of Sheppey and Grain, after it has received the water
of the Swale.

The Ravensbourne, a little stream ten miles long,
rises near Keston Common and flows past Bromley and
Lewisham to Deptford. The Cray, which gives its
name to several villages and to Crayford, rises near
Orpington and falls into the Darent near Dartford.
The Darent, which gives its name to Darenth and
Dartford, rises near Westerham and runs under the
North Downs, through which it forms a pass. It passes
Otford, Eynesford, and Sutton on its way to Dartford,
and after a course of twenty miles falls into the Thames.
The country through which the Dart flows is very pic-
turesque, and, in part of its course, it provides water-power
for paper mills and gunpowder works.

Besides the rivers flowing into the Thames, there are

the Stour, which flows in a north-easterly direction to the Strait of Dover, and the Rother, which is the boundary on the Sussex side. The Stour is entirely a Kentish river throughout its course of 40 miles. It rises in two streams, which are known as the Greater and the Lesser Stour. The Lesser Stour rises near Lyminge and joins

Reculver Church

the other river at Stourmouth. The Greater Stour, or main stream, rises near Lenham, and passing Ashford and Canterbury, proceeds to its junction with the Lesser Stour. The two rivers now form a channel, flowing by two mouths to the sea, so as to form the Isle of Thanet. The northern branch falls into the sea at Reculver, the

southern passes Sandwich to enter Pegwell Bay. This
channel, formerly called the Wantsum, was at one time
of great importance. Both the Greater and Lesser Stour
have excellent trout and salmon-trout.

The Rother rises at Rotherfield in Sussex, and for
several miles forms the boundary between Kent and
Sussex. It enters the sea at Rye, but formerly its mouth
was at New Romney. In the reign of Edward I it
deserted the old channel for the present one.

6.   Geology and Soil.

By Geology we mean the study of the rocks, and we
must at the outset explain that the term *rock* is used by
the geologist without any reference to the hardness or
compactness of the material to which the name is applied;
thus he speaks of loose sand as a rock equally with a hard
substance like granite.

Rocks are of two kinds, (1) those laid down mostly
under water, (2) those due to the action of fire.

The first kind may be compared to sheets of paper
one over the other. These sheets are called *beds*, and such
beds are usually formed of sand (often containing pebbles),
mud or clay, and limestone, or mixtures of these materials.
They are laid down as flat or nearly flat sheets, but may
afterwards be tilted as the result of movement of the
earth's crust, just as you may tilt sheets of paper, folding
them into arches and troughs, by pressing them at either
end. Again, we may find the tops of the folds so pro-

duced washed away as the result of the wearing action of
rivers, glaciers and sea-waves upon them, as you might
cut off the tops of the folds of the paper with a pair of
shears.   This has happened with the ancient beds forming
parts of the earth's crust, and we therefore often find them
tilted, with the upper parts removed.

The other kind of rocks are known as igneous rocks,
which have been melted under the action of fire and
become solid on cooling.   When in the molten state
they have been poured out at the surface as the lava of
volcanoes, or have been forced into other rocks and cooled
in the cracks and other places of weakness.   Much
material is also thrown out of volcanoes as volcanic ash
and dust, and is piled up on the sides of the volcano.
Such ashy material may be arranged in beds, so that it
partakes to some extent of the qualities of the two great
rock groups.

The production of beds is of great importance to
geologists, for by means of these beds we can classify the
rocks according to age.   If we take two sheets of paper,
and lay one on the top of the other on a table, the upper
one has been laid down after the other.   Similarly with
two beds, the upper is also the newer, and the newer will
remain on the top after earth-movements, save in very
exceptional cases which need not be regarded by us here,
and for general purposes we may regard any bed or set of
beds resting on any other in our own country as being
the newer bed or set.

The movements which affect beds may occur at
different times.   One set of beds may be laid down flat,

then thrown into folds by movement, the tops of the beds worn off, and another set of beds laid down upon the worn surface of the older beds, the edges of which will abut against the oldest of the new set of flatly deposited beds, which latter may in turn undergo disturbance and renewal of their upper portions.

Again, after the formation of the beds many changes may occur in them. They may become hardened, pebble-beds being changed into conglomerates, sands into sandstones, muds and clays into mudstones and shales, soft deposits of lime into limestone, and loose volcanic ashes into exceedingly hard rocks. They may also become cracked, and the cracks are often very regular, running in two directions at right angles one to the other. Such cracks are known as *joints*, and the joints are very important in affecting the physical geography of a district. Then, as the result of great pressure applied sideways, the rocks may be so changed that they can be split into thin slabs, which usually, though not necessarily, split along planes standing at high angles to the horizontal. Rocks affected in this way are known as *slates*.

If we could flatten out all the beds of England, and arrange them one over the other and bore a shaft through them, we should see them on the sides of the shaft, the newest appearing at the top and the oldest at the bottom, as shown on p. 24. Such a shaft would have a depth of between 10,000 and 20,000 feet. The strata beds are divided into three great groups called Primary or Palaeozoic, Secondary or Mesozoic, and Tertiary or Cainozoic, and below the Primary rocks are the oldest rocks of Britain, which form as it were the foundation stones on which the other rocks rest. These may be spoken of as the Precambrian rocks. The three great groups are divided into minor divisions known as systems. The names of these systems are arranged in order in the figure with a very rough indication of their relative importance, though the divisions above the Eocene are made too thick, as otherwise they would hardly show in the figure. On the right hand side, the general characters of the rocks of each system are stated.

With these preliminary remarks we may now proceed to a brief account of the geology of the county.

In considering the geology of Kent it will be well, at the outset, to note that, as a rule, the older rocks are exposed on the surface in the south of the county, while the more recent are found in the north. Further, it is well to remember that two great forces have helped to give Kent its present physical form. First, the district known as the Weald has been raised, while the beds in the north of Kent have been depressed. The second process has taken place by the agency of rain and rivers, so that the softer parts have disappeared more rapidly than the hard. This accounts for the presence of the hills and valleys of Kent.

We will begin with a description of the oldest rocks, and proceed step by step to the newer and later-formed rocks. The Wealden Beds, which occur in the south of the county, have a thickness of over 2000 feet. The strata have been divided into two classes, the Hastings Beds and the Weald Clay. It is believed that the entire mass was deposited in the estuary of a great river, for the fossils found are either fresh-water shells, or bones of crocodiles or of a great land reptile.

The Hastings Beds are chiefly sandy, and are in the extreme south of the county. The scenery of this tract is varied and beautiful, and the ground is hilly. Very fine exposures of the sandstone rocks are to be seen round Tunbridge Wells, where Harrison's Rocks and the Toad Rock at Rusthall Common are the best examples.

The Toad Rock, near Tunbridge Wells

The Weald Clay is a low, flat tract of land from four to six miles wide, which runs east and west by Tunbridge to Romney Marsh. It is a brown or blue clay, badly drained and mostly in pasture. At Hythe the Weald Clay yields fresh-water and marine shells, showing a depression of the estuary and the beginning of marine conditions.

Above the Weald Clay comes the Lower Greensand, a hilly ridge running from Hythe in a north-westerly direction to Sevenoaks and then west to Westerham. There are several subdivisions of the Lower Greensand, but the most important is known as the Hythe Beds. They contain valuable beds of limestone known as Kentish Rag, which are largely quarried at Maidstone. This stone is in great demand for building churches and other important buildings in London and its suburbs.

We now come to the Gault, which occurs between the Lower Greensand and the Chalk. It forms a narrow valley, the bottom of which is stiff blue clay. The Gault may be well seen at Folkestone, where, at Copt Point, a section of the bed is exposed over the Lower Greensand. It contains many fossils, but the shells are difficult to preserve. The Upper Greensand is very thin in Kent, but near Folkestone some greenish sandy beds about twenty feet thick are seen lying above the Gault.

The Chalk succeeds the last formation and slopes gently to the north, but abruptly to the south. Its thickness varies from 600 to 700 feet, and it contains some of the highest hills in Kent. Chalk is a white, soft, pure limestone composed of countless shells or foraminifera. It was probably formed in a deep, open sea. The Chalk is found both with and without flints. To the north of Maidstone the latter kind is worked, and after it is ground up and mixed with clay it forms Portland cement.

	Names of Systems		Characters of Rocks
TERTIARY	Recent & Pleistocene Pliocene Eocene		sands, superficial deposits clays and sands chiefly
SECONDARY	Cretaceous		chalk at top sandstones, mud and clays below
	Jurassic		shales, sandstones and oolitic limestones
	Triassic		red sandstones and marls, gypsum and salt
PRIMARY	Permian		red sandstones & magnesian limestone
	Carboniferous		sandstones, shales and coals at top sandstones in middle limestone and shales below
	Devonian		red sandstones, shales, slates and limestones
	Silurian		sandstones and shales thin limestones
	Ordovician		shales, slates, sandstones and thin limestones
	Cambrian		slates and sandstones
	Pre-Cambrian		sandstones, slates and volcanic rocks

The Chalk ends the series of strata known as the Secondary Rocks, which are succeeded by Tertiary formations of a recent period. They are found in a tract of land, from six to eight miles wide, running along the south side of the Thames. The oldest of the Tertiaries are the Thanet Sands, which are evident at Pegwell Bay and Reculver. They are fine grey or buff sands, but quite destitute of fossils.

The Woolwich Beds rest on the Thanet Sands, and are to be seen in the great pits at Lewisham and Charlton. They are also exposed at Cobham Park, near Canterbury, and at Herne Bay. They consist of mottled clays and sand and pebbles, in which some fossils are found.

We have now reached the London Clay, which is 480 feet thick in the Isle of Sheppey. It forms the top of Shooter's Hill and is found in the wooded tract between Canterbury, Whitstable, and Herne Bay. In Sheppey it contains bands of septaria, which are nodules of carbonate of lime and iron pyrites. The former is used to make cement, and the latter to make copperas. The London Clay is also used for brick and tile making. In the London Clay great numbers of fossils have been found, such as nautilus shells, crabs, turtles, sharks, birds, and over 200 species of plants.

The last of the geological divisions of Kent we shall class as the Alluvial deposits. These are found in a narrow strip along the Thames, at Woolwich, Plumstead and elsewhere; on either side of the Medway estuary; and in the district of the Romney Marsh.

Before we pass to the next chapter, we may profitably

consider the soil of the county. Kent is one of the chief
agricultural counties of England, and has long been
famous for the fertility of its soil, and the varied and
bountiful crops it produces. In such a large county,
and with such different physical features, the soil varies
in quality and character. Chalk soils of little value are
found on the sides of hills and on the borders of the
Thames. Loamy soils of various depths and quality
are common in the valleys. Clay soils that are cold
and tenacious are common in the west of Kent, or
stiff and heavy as in parts of the Weald. Gravelly soils
prevail about Dartford and Blackheath, while sandy soils
are frequent in West Kent, on the commons and heaths.

Sheppey has much deep, strong, stiff clay, and its
marshes have a surface of vegetable mould. Romney
Marsh has a soil of fine soft loam, with a mixture of
sea-sand. Thanet is in a high state of cultivation, owing
to its rich soil; while the Weald has long been famed
for its fertility.

7. Natural History.

Various facts, which can only be shortly mentioned
here, go to show that the British Isles have not existed
as such, and separated from the Continent, for any great
length of geological time. Around our coasts, for instance,
are in several places remains of forests now sunk beneath
the sea, and only to be seen at extreme low water.
Between England and the Continent the sea is very
shallow, but a little west of Ireland we soon come to

very deep soundings. Great Britain and Ireland were thus once part of the Continent, and are examples of what geologists call recent continental islands. But we also have no less certain proof that at some anterior period they were almost entirely submerged. The fauna and flora thus being destroyed, the land would have to be restocked with animals and plants from the Continent when union again took place, the influx of course coming

The Penshurst Oak

from the east and south. As, however, it was not long before separation occurred, not all the continental species could establish themselves. We should thus expect to find that the parts in the neighbourhood of the continent were richer in species and those furthest off poorer, and this proves to be the case both in plants and animals. While Britain has fewer species than France or Belgium, Ireland has still less than Britain.

The flora of Kent is as rich and varied as that of any part of England. Indeed, Kent would not be called the Garden of England if it were not well stocked with flowers. This abundant and diversified flora is partly owing to the fact that Kent possesses the necessary conditions to enable nearly all our plants to find a suitable home.

The county has a long stretch of tidal river with its fresh-water and brackish marshes. There are cliffs of stiff London clay between Whitstable and Reculver, which are followed by the chalk cliffs of Thanet, and the sandy shingle of Pegwell Bay. Extensive sand-dunes near Sandwich, Deal, and New Romney; the sheltered undercliff from Dover to Folkestone; and the coarse shingly beach at Dungeness together afford a variety of soil for the most varied flora around the long coast.

The inland parts offer equally varied conditions. The different soils, geological strata, elevation, and exposure to shade and moisture favour great diversity in the flora. The most important and distinctive feature of the Kentish flora is to be found in the abundance and variety of the orchidaceous plants. Out of 44 British species, no less than 33 are found in Kent. The purple orchis and the spotted orchis are especially common.

Lord Avebury, who is better known as Sir John Lubbock, has written a charming book on *The Scenery of England*. In that volume he gives a botanical sketch of the Kentish commons; and it will be interesting to read his description of them : "A Kentish common is, however, no mere bit of bare worthless land, sparsely covered with bents and other coarse grasses and weeds, but is set with birches and junipers, broom and gorse, wild roses and hollies, yews and guelder roses, clematis and honeysuckle, growing over white, pink and blue milkwort, blue veronica, pink heather, and yellow rock-rose, sweet with the fragrance of the furze and roses, and the aromatic scent of the pine woods. In the hollows are many pools, fringed by reeds and rushes, irises and water grasses, with green carpets of sphagnum, studded with red sundew, and dotted over with the pure white flossy flags of cotton grass; while on the water repose the beautiful leaves and still more lovely flowers of the lilies, over which hover many butterflies, while brilliant metallic dragon-flies flash or dart about."

The number of species of animals found in Kent is fairly large, so that it will be possible to refer to only a very few of them. Reference will be made to some of the domestic animals in the chapter on agriculture, and the fishes will be noticed in the chapter on the fisheries of Kent.

The wild animals of Kent are similar to those that are found in most English counties. They comprise among others the badger, fox, hare, rabbit, squirrel, stoat and weasel. The otter is almost extinct. Fallow deer are preserved for the parks; and the fox is kept for the chase.

We cannot look in Kent for any great abundance of the northern water-fowl or larger birds of prey, such as we find, for example, on the firths and moors of Scotland or northern England, but taken as a whole the birds of

Kent are of very numerous species—as numerous, probably, as those of any other county. This is partly due
to the varied nature of the country, to which we have
already referred, and partly to geographical position.
Kent, it is true, does not experience the full force of
the stream of migration which strikes Lincolnshire and
Norfolk with such especial strength in the season, but,
on the other hand, she receives—as does the neighbouring
county of Sussex—many a rare continental visitor from
the south, and hence brilliantly coloured strangers like
the bee-eater or golden oriole are more frequently seen
here than further north (too often, unhappily, to fall
victims to the ruthless collector) and many a rare warbler
escapes notice by virtue of its inconspicuous plumage.
Kent is especially noteworthy as having three birds particularly appertaining to it so far as their names are
concerned, though they are not in any way limited to the
county. They are the Kentish plover, the Sandwich tern,
and the Dartford warbler. The two former are found over
a large part of the Old World, and the latter in various
places in southern Europe.

Just as we noticed that the woods, the marshes,
the shingle, and the sand accounted for the luxuriant
flora of Kent, so we may observe that the same varieties
of soil and flora account for the butterflies and moths and
the numerous insects. It is well known that many larvae
thrive mainly, if not solely, on their own special food-
plants, and thus, if a certain species is required it is only
necessary to cultivate the food-plant, and the desired
species will soon appear. Thus we need not be surprised

that the extensive flora of Kent offers every attraction to
species that we might expect to find.

8. Round the Coast—London to Shep-pey.

The coast-line of Kent is of so much interest that we
shall want several chapters to describe it. The river-
coast begins at Earl's Sluice, a little above Deptford.
From this point to Sheerness, where the estuary of the
Thames may be said to end, is a distance of forty miles.
The shore-line of the Medway estuary is about twelve
miles. The sea-coast, which is measured from Sheerness
to the point where Kent and Sussex meet, is upwards of
ninety miles. We thus get a total length of water-line
along river and sea measuring upwards of 140 miles.

We will now take an imaginary trip along the Kentish
shore of the Thames from Earl's Sluice to Sheerness.
Earl's Sluice, a little below the Commercial Docks, divides
Surrey from Kent. Just beyond, and four miles from
London Bridge, is Deptford. In the time of Henry VIII
Deptford was a royal dockyard; and it thus continued
for a long time. As a dockyard it has ceased to exist,
and part of the site is now occupied by the Foreign Cattle
Market. It was at Deptford on April 4, 1581, that
Queen Elizabeth visited the *Golden Hind*, the ship in
which Drake had "compassed the world." The Queen
dined on board, and after dinner knighted the gallant
Drake.

Below Deptford, we come to Greenwich, the "Green-town," which was a favourite station of the old North-men. Greenwich is a town that has played an important part in our history, and as we shall make further reference to this "royal" borough, we will only note that it is famous for the Hospital and the Observatory, both of which are of national importance.

Greenwich Hospital

Leaving Greenwich, we soon reach Woolwich, which once had a dockyard, but is now famous for the Royal Arsenal. On the opposite shore is North Woolwich, which has a steam-ferry running every few minutes. Among the historic ships that have been launched from Woolwich dockyard we may mention the *Queen Elizabeth*, in 1559, the *Royal Sovereign*, in 1637, and the ill-fated *Royal George*, in 1751. It was about this last vessel that Cowper wrote his poem "The Loss of the *Royal George*."

At Woolwich, the river is a quarter-of-a-mile wide, and the land on the Kentish shore begins to rise. Erith, "the old haven," is about four miles distant, and is rapidly becoming a riverside town of some note. Near Erith pier there are some public gardens along the river, and about a mile lower we come to Dartford Creek. The river now runs south-east until we reach Greenhithe, which has considerable trade in chalk and lime. From Greenhithe, in May, 1845, the *Erebus* and *Terror*, under Sir John Franklin, sailed on their last fatal expedition to the Polar Seas. Off Greenhithe lies the *Worcester*, a ship which is used as a training college to prepare the sons of gentlemen for the naval profession. The river now turns north-east and then south-east to Northfleet, which is quite close to Gravesend.

Much chalk is still burnt at Northfleet, and lime is exported to Holland and elsewhere. The flints from the chalk-pits are sent not only to Staffordshire, but even to China, for the use of the potteries. Gravesend almost forms one town with Northfleet and is a place of considerable importance, for it occupies the first rising ground after entering the river. Outward-bound vessels lie here to complete their cargoes; and the right to convey passengers to and from London was once the privilege of the boatmen of this town. Sebastian Cabot in 1553, and Martin Frobisher in 1576, assembled their little ships at Gravesend, after Queen Elizabeth had wished them farewell at Greenwich.

Below Gravesend the river widens to half-a-mile and the depth at low water is 48 feet. Here the Thames forms *The Hope*, the last of its sixteen reaches. The banks on either side are flat, and there is little of interest, for no object breaks the level line of shore. After passing Hope Point, the river runs almost due east to the Isle of Grain, and here its waters mingle with those of the Medway. On the Isle of Grain is Port Victoria, whence steamers run to Sheerness and the Continent. Sheerness is a dockyard on the Island of Sheppey, and of great importance, as it is the headquarters of the Commander-in-Chief at the Nore.

The Nore Lightship is on the Nore Sand, midway between Sheerness and Shoeburyness, where the Thames is six miles wide. We are now fairly in the North Sea and the long line of the Kentish coast runs south-east to Warden Point.

9. Round the Coast —— Sheppey to Dungeness.

The Swale is the channel, or river as it is sometimes called, which separates Sheppey from the mainland. It was formerly part of the water-way from London to Dover, but has long ago lost its importance. Crossing the Swale, we arrive at Whitstable, an ancient town, famous for its oyster fisheries. This picturesque, yet untidy port is full of queer nooks and unexpected corners. Between Whitstable and the opposite Essex coast, the tideway has a breadth of 18 miles. From Whitstable onwards for nearly 20 miles the coast runs generally east with not a harbour of importance.

Margate

3—2

Herne Bay is a rapidly growing sea-side town, a few miles east of Whitstable. It is a popular resort in the season, and, among its attractions, there are good sands and a long pier. The cliffs become higher as we approach Reculver, which has only twin spires to remind us of its former importance, when it commanded the northern entrance to the river Wantsum, separating Thanet from the mainland. The Wantsum was once used by vessels to avoid the passage round the Forelands.

Birchington is next reached; there the chalk cliffs, which are such a prominent feature on the Kentish coast, are first seen. From this town, they continue almost without a break to Ramsgate.

Westgate is next passed on our way to Margate, which is one of the most popular of all the Kentish seaside resorts. Londoners have long valued a trip to Margate, "Gate of the Sea," and every year it seems to offer greater attractions to its visitors. We now round the Isle of Thanet, passing the North Foreland, on our way to Broadstairs and Ramsgate. The former is much frequented for its quiet, its bracing air, and extensive sands. Upon a point of the cliff overlooking the harbour and pier is Bleak House, which was, for some time, the residence of Charles Dickens.

Ramsgate has a south-western aspect and is consequently milder than Margate. It has a harbour of more than 50 acres in area and capable of receiving 400 vessels at one time. It thus forms a harbour of refuge for the Downs, and, with its two fine piers, is a very favourite watering place. Ramsgate is near Pegwell Bay, famous

The Fisher Gate, Sandwich

for its shrimps.   Ebbsfleet is of historic interest, for it was probably here that Hengest and Horsa landed in the fifth century and St Augustine and his missionaries at the close of the sixth century.   Further south is Sandwich, a decayed seaport, now famous for its golf-links.

Leaving Sandwich, we find that the coast runs almost due south to the South Foreland.   From Sandwich to Deal the coast is generally low, marshy, and uninteresting. Deal and Walmer have two historic castles and fine beaches.   Looking seawards at Deal, the Downs present an animated scene with their numerous vessels.   When St Margaret's is reached the cliffs are high and there is a considerable bank of shingle.   There is a lighthouse at the South Foreland, which stands 374 feet above sea-level.   From the cliffs, a magnificent view can be obtained across the Strait of Dover, which is here at its narrowest.   Between St Margaret's Bay and Dover the chalk cliffs attain a height of from 300 to 400 feet.

Dover has always been an important port from the time of the Romans.   Between the cliffs, we see a busy port, with its fine harbour and piers; while on the heights are the fortifications and barracks.   Southward from Dover is the celebrated Shakespeare Cliff, 350 feet above the sea.   Near this spot, it is contemplated to construct a tunnel to Calais.

Between Dover and Folkestone the cliffs consist of chalk and rise to 500 feet.   In East Wear Bay a great landslip took place in the middle of the last century and formed an undercliff known as the Warren—a picturesque tract and a happy hunting ground for the geologist and botanist. Folkestone is a fashionable seaside resort, and has increased in importance owing to the growing trade with the Continent. Beyond Folkestone, the cliffs get lower, and the coast makes a bold curve to the south-west to Dungeness. Sandgate and Hythe adjoin Folkestone, and as they are on a level with the sea, constant inroads are being made on the coast.

Beyond Hythe, the coast borders on the district known as Romney Marsh, and has to be protected by a sea-wall, of which some account will be given in another chapter. Passing New Romney at the end of Dymchurch Wall we soon arrive at Dungeness, where there is one of the most remarkable collections of shingle to be found on the English coast.

Dungeness is a dangerous headland for ships to weather, and the neighbouring sea has been the scene of many shipwrecks. It was off Dungeness, on January 22, 1873, that the *Northfleet* was run down in a fog with a loss of 300 lives. From Dungeness, the coast bends to the west, and for a few miles nothing of interest is seen till the coast of Sussex is reached.

10. The Coast—Gains and Losses.

Our English coasts are remarkable for many changes during the last eight or nine hundred years. In some parts the sea has retreated, while elsewhere the land has yielded to the ocean. There are few parts of our coast that show so many changes as those made in the coast of

Kent. Along the Thames estuary and thence as far as Folkestone the sea has made serious inroads on the land; but from Folkestone to Dungeness the land has gained at the expense of the sea.

Folkestone Harbour

The most serious losses of land have occurred on Sheppey, at Herne Bay, Reculver, and Whitstable; while the sea has retreated from the following towns that were once seaports—Lymne, Romney, Hythe, Richborough, Stonar, Sandwich, and Sarre. Thanet, that was once an island, is now only so by name, for the Wantsum is practically dried up. The Swale, which divides Sheppey from the mainland, is now only a portion of its former width.

Let us notice the portions of Kent that have suffered most from the inroads of the sea. The cliffs on the north of Sheppey are being destroyed at a rapid rate, at least 50 acres having been lost in 20 years. The church at Minster, now near the coast, was in the middle of the island in 1780. If the present rate of destruction continues, the whole island will be lost at a not very remote date.

Herne Bay can no longer be called a bay, for the sea has washed away the former headlands, so that the line of coast is almost straight. The twin spires of Reculver mark the site of an ancient town bearing that name. Reculver was once a flourishing Roman city, having its castle and camp, and a mint for the coinage of gold and silver. Not only did Roman emperors live here, but the early kings of Kent made it their home.

The loss of Reculver has been very gradual. In Leland's time it was half-a-mile from the sea. In 1785, the north wall of the castle had been lost by a fall of the cliff. The churchyard was entire in 1805, and there was a highway between the church and the cliffs, but in 1809, the distance of the church from the cliffs was only a few yards.

An eminent geologist has reckoned that the cliffs between Reculver and the North Foreland lose about two feet yearly; while those on the south of Thanet, between Ramsgate and Pegwell Bay, lose as much as three feet each year.

Shakespeare Cliff at Dover has lost much of its grandeur since Shakespeare's time. There was a great landslip from this cliff in 1810, and one yet greater in 1872; while quite recently other falls from this lofty eminence have occurred. There is every reason to believe that Dover itself has suffered much from the sea, for the harbour was, in former times, an estuary.

The famous Goodwin Sands are said to have formed part of the mainland in Earl Godwine's days. According to history this portion of Godwine's estate was overwhelmed by the sea in 1099. Shakespeare refers to the Goodwin Sands in the *Merchant of Venice* as follows:—

" The Goodwins, I think they call the place; a very dangerous flat and fatal, where the carcases of many a tall ship lie buried."

These sandbanks consist of about 15 feet of sand resting on blue clay. They are divided into two parts, the North Goodwin and the South Goodwin; and between them is an inlet called Trinity Bay. At high water, the sands are covered, but at low water they may be walked upon with safety, and games of cricket have been played on them. We shall refer to the Goodwin Sands when we deal with the lighthouses and lightships round the Kentish coast.

The most extensive district reclaimed from the sea is that known as Romney Marsh. The work of reclamation was probably begun by the Romans, and continued by the Saxons. The first portion to be reclaimed was an island, on which the town of Old Romney now stands. The old town of Lydd once stood upon another island, as

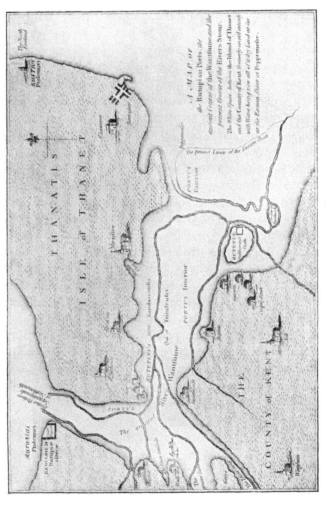

Thanet and the River Wantsum (from an old map)

did Ivychurch, Old Winchelsea, and Guildford. The sea swept round them, and rose far inland at every tide. Burmarsh and other districts were reclaimed more recently, and by degrees the whole became firm land.

Large additions were made to it from time to time by the deposits of shingle along the coast, which left several towns, formerly seaports, stranded upon the beach far inland. Lymne is left high and dry more than three miles from the sea, and sheep now graze where the Roman galleys once were rowed. West Hythe, once a Cinque Port, has now a wide stretch of shingle between the town and the sea. Old Romney, past which the Rother flowed, is about two miles from the sea.

Dungeness, running almost due south, gains accumulations of shingle so rapidly, that it is said to have extended more than a mile seaward within the memory of persons now living. This deposit extends from the shore in the form of a triangular promontory, the base of which measures six miles and the length three miles. The shingle deposit covers an area of 6000 acres, having a height of about five feet above high water. The bulk of the pebbles consists of flint, and mixed with these are chert and other stones from the Wealden series and Greensand formation.

11. The Protection of the Coast—Sea-walls and Groynes.

We will now consider some of the means that have been adopted to save the coast from further incursions of the river and the sea. Let us begin with the walls and embankments along the Thames. Before the river had been confined within its present channel, it was a very much broader estuary. In many parts, between London and Gravesend, it was several miles wide. The higher tides covered Plumstead and Erith marshes on the south, while the East Ham and Barking marshes were also submerged on the north, or Essex shore. Thus the river meandered in many winding channels at low water, leaving on either side expanses of rich mud and ooze.

It is not certain whether the Britons or the Romans built the walls of the Thames, but they are the result of skill and bold enterprise. The river is now several feet higher than the level of the surrounding country, and is practically an aqueduct, raised and supported between its artificial banks. It has been estimated that these embankments, winding along the river side, up creeks and little streams, round islands and about marshes, from London to the mouth of the Thames, are not less than 300 miles long.

At various periods breaches in the walls of the Thames have occurred, and their repair has often been most difficult and expensive. In the reign of Henry VIII

the marshes at Plumstead and Lesnes were submerged
and were not reclaimed for a long time; and the low
lands east of Greenwich were also inundated and reclaimed at a later period. From this time the banks
on the south side of the Thames have been secured
from breaches.

Further along the estuary of the Thames, sea-walls
have been built at Westgate, Margate, and elsewhere.
Between Birchington and Herne Bay the sea-wall and
coast have been protected by groynes. These groynes
consist of oak posts, 10 feet long, and spaced from 4 to
5 feet apart, driven into the clay beach. To the posts
are bolted strong horizontal planks. The groynes extend
at right angles from the shore, and their length is about
130 feet. In front of Herne Bay there are 91 groynes,
and all along the sea-coast a similar system of groynes
prevails.

These groynes have been the means of stopping the
drift of the shingle to the west, and causing a great accu-
mulation above high-water level. Most of the authorities
along the coast forbid the carting away of shingle, which
is now of such importance in protecting the cliffs from
the wasting effects of the sea.

On the south-east coast of Kent, and north-east of
Dungeness, is a low flat tract of valuable grazing land,
which is protected from the sea by Dymchurch Wall.
This extensive district is known as Romney Marsh, and
its surface is from 8 to 10 feet below high-water level.
It is therefore certain that, if this artificial protection
were removed, the sea would quickly submerge the

whole district, that was reclaimed at such trouble and expense. Dymchurch sea-wall, which extends from New Romney to Hythe, a distance of 4 miles, was constructed by the Romans. The top of this bank is 20 feet wide, and from 10 to 13 feet above high water. The sea breaks against the wall over its entire length, and great damage is done by south-east gales, which blow dead on shore.

Romney Marsh

From Hythe to Folkestone it has been necessary to construct sea-walls and promenades, and place a number of high groynes for the protection of the coast. As recently as 1899 and 1900 heavy gales broke with great violence against the sea-wall at Sandgate, where a breach of 40 feet long was made, and thousands of tons of shingle were thrown on the roadway.

12. The Coast—Sand=banks and Light= houses.

The navigation of the Thames is now very difficult, and it requires great skill on the part of pilots to avoid the numerous shoals and sand-banks that occur in the estuary. A glance at a good map of the Thames will show that the navigable portion of the river is a narrow channel in the middle of the stream. Some of the sand-banks on the Kentish side are as follows :—Nore Sand, The Flats, and Margate Sand.

Off the east coast of Kent are the Goodwin Sands, which have long been of evil repute to mariners. These famous sands extend for 10 miles, at a distance of 5 or 6 miles from the coast.

Between the Goodwins and the coast there is a safe anchorage for ships, which is known as the Downs. The water varies in depth from 4 to 12 fathoms, and it is the largest natural harbour of refuge round our coast.

Now in order to assist mariners in navigating our river-mouths and shores there are placed, at various points along the coast, lighthouses, lightships, beacons and buoys. It is a curious fact that very little was done in the way of lighting our coasts till the beginning

The Pharos, Dover

of the nineteenth century. It is true that about 2000 years ago towers or beacons were erected on the coast by the Romans, and one of these, known as the Pharos, still stands at Dover. But it is worthy of note that a hundred years ago there were only about 30 lighthouses and lightships round the British coasts, whereas now there are nearly 900.

The authority that has "the duty of erecting and maintaining lighthouses and other marks and signs of the sea" is known as the Elder Brethren of Trinity House. They derive their income from light-dues levied on shipping, and as this amounts to £300,000 per annum, they are able to erect and maintain lighthouses, lightships, beacons, buoys, &c. They also have the power to appoint and license pilots, and remove wrecks when dangerous to navigation.

Now let us glance at a few of the lightships and lighthouses around the Kentish coast. In the estuary of the Thames is the Nore Sand, and on it is fixed the famous lightship which guides the shipping in and out of the Port of London. This light was first placed here in 1732 by Mr Hamblin, who moored a vessel called the *Experiment* on this sand. The light from this ship was found so valuable that the Nore Light was placed under the control of Trinity House. The vessel was a red hull, with the name *Nore* on its sides. The white light revolves every half-minute.

Rounding the Isle of Thanet, midway between Margate and Ramsgate, there is the North Foreland Lighthouse. It is one of the oldest in England, for it was first erected in 1683, to take the place of a beacon that had long warned mariners off the Goodwin Sands. It was rebuilt in 1790, and altered in 1880. The present light is visible for 20 miles, and occults every half-minute, showing white and red for twenty-five seconds, followed by an eclipse of five seconds.

As we should expect, we find that the Goodwin Sands are well lighted. There are four lightships, and in foggy weather fog-horns and fog-sirens are sounded. Besides the lightships, no less than nine buoys are moored round the Sands. The Goodwin lightships are as follows: the North Goodwin Lightship, which shows a white flash-light; the South Goodwin Lightship, which shows a double-flash white light; the Gull Lightship, which displays a white revolving light; the East Goodwin Lightship, which shows a green revolving light. These four lights are visible at a distance of 10 or 11 miles in clear weather. Each vessel has a red hull, with its name on its sides.

At the South Foreland there is now only one lighthouse at a height of 374 feet. It shows a white flash-light visible at a distance of 26 miles.

The Point of Dungeness is most dangerous, and very difficult for ships to weather. At one time, there were no fewer than 20 wrecks visible in the East Bay. The main lighthouse is about 310 yards within high water mark, and is a blue brick tower. Its white flash-light is visible at a distance of 17 miles. A fog-siren is sounded in thick weather, and there is also a signalling station. The other lighthouse, with a red cylindrical tower, stands 38 feet above high water and 480 yards from the main lighthouse.

13. Climate and Rainfall.

When we speak of the climate of a district we mean the average weather of that district. The climate of our country is everywhere changeable, yet the average of the changes is not the same for all parts. Thus we should say that Kent is on the whole drier and colder than Cornwall, and if we enquired why this is the case, we should find that there are many reasons for this difference. Situation, elevation, vegetation, and many other minor causes are at work to make the climates of Kent and Cornwall different.

Again, even in Kent itself we find there is variety in the climate. The west is drier than the east, and the north is colder than the south. In other districts we find local influences at work, such as configuration and aspect, which tend to make a difference in the climate of two places not far apart. A hill-slope facing south-wards receives the rays of the sun more directly than does a slope towards the north. Thus it comes about that the southern faces of hills are more sunny and more genial. We also know that a sandy surface is the cause of greater extremes of temperature than a heavy clay soil ; and that vegetation tends to make the climate more equable.

The climate of a country has a marked influence on its productions, so that we find crops grown in the south-east of England that would not grow in the northern part. We know, too, how the climate in-

fluences the character of people, and how the weather
is one of the factors in our enjoyment of life. There
is a Meteorological Society in London which collects
from numerous stations particulars of the temperature
of the air, the hours of sunshine, the rainfall, and the
direction of the winds. The summary of these appears
day by day in our newspapers, so that we see at a
glance, by means of a chart or map, exactly what kind
of weather has been experienced over the British Isles.
At the end of the year these results are totalled and
averaged, and we are then in a position to compare and
contrast the character of the climate at various places.
In this chapter we will compare some of the climatic
results of Kent with those of England or Great Britain,
as the case may be.

In 1905 the mean temperature of England was 48·7°,
while that of Kent was 49·5°. The mean temperature
in Kent varied from 50·5° at Margate to 48·7° at Tunbridge Wells. This is a considerable difference, but can
be easily explained by the nearness of Margate to the sea,
and the fact that Tunbridge Wells is inland.

The hours of bright sunshine in England in 1905
amounted to 1535, and in Kent to 1667·8. Here, again,
Kent is in advance of the average for the whole country ;
but, curiously, Tunbridge Wells, for the same year, had
1712·4 hours of bright sunshine against 1577·3 at Margate.
Margate had 74 sunless days and Tunbridge Wells 60
sunless days.

The information with regard to the rainfall of a
country is of the highest importance, and it will be

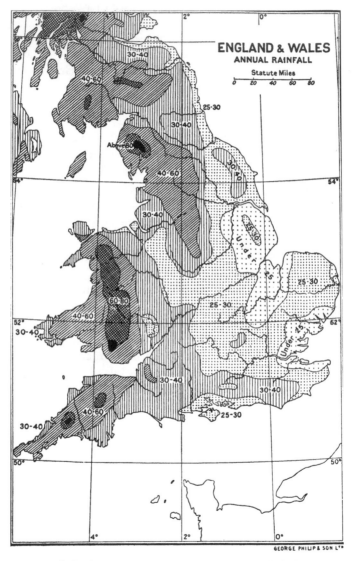

ENGLAND & WALES
ANNUAL RAINFALL

Statute Miles

(*The figures express the annual rainfall in inches*)

true to say that the yearly rainfall in England and Wales decreases from west to east. The highest rainfall recorded in England and Wales, in 1905, was at Glas Lyn, near Snowdon. Here no less than 176·6 inches were mea-

Greenwich Observatory

sured, while at Shoeburyness, in Essex, the lowest rainfall of 14·57 inches was measured.

Now both of these records are extremes. The average rainfall for Great Britain in 1905 was 27·17 inches, and

of Kent 25·15 inches. The highest rainfall in Kent in
that year was at Paddlesworth, with 43·57 inches, and
the lowest was at Herne Bay, with a record of 18·32.
It is also interesting to note that in 1905 there were
186 rain days in Great Britain, and 178 in Kent. A
rain day is one on which 0·005 inch or more is measured. Greenwich is a most important station for
meteorological observations. Its rainfall for 1905 was
23·024 inches; its rain days numbered 161; and its
wettest month was June, when 4·323 inches of rain fell.

Now let us compare the rainfall of Margate and
Tunbridge Wells. In 1905 the rainfall at Margate
was 21·18 inches, which fell on 173 days. At Tunbridge Wells the rainfall was 27·05 inches on 177 days.
Of course these results vary from year to year, but they
have been collected for a number of years, so that the
average may easily be worked out for any place that has
its station for the collection of weather data.

Land fogs are now rare along the south of Kent,
owing to improved drainage. There are sea-fogs in the
east, where the Atlantic and Arctic waves of temperature
are apt to clash. In the neighbourhood of London and
along the Thames there are frequent fogs in the autumn
and winter.

The prevailing winds are southerly and south-westerly,
and from them there is least protection. The south of
Kent is, however, well protected from the northerly and
easterly winds. The high land of the Downs is remarkable for its brisk bracing air ; and the sea-side resorts all
round the coast have a bracing climate.

14. People—Race, Dialect, Settlements, Population.

Long before our English forefathers settled in Kent, this district was inhabited by a British tribe known as the Cantii. We know something of these Britons from a traveller named Pytheas, who visited Kent before Caesar's invasion. The Cantii resembled the people of Gaul in habits and mode of life, and probably there was some connexion between the people on both sides of the Channel.

After the Romans had defeated the Britons, the work of civilisation went on, until the victors had to leave our shores, and the Northmen settled in our land. In 449 A.D. the Jutes landed in Kent, which they eventually conquered, for they either killed the Britons or drove them away to the west. Thus the history of the county of Kent, as far as its people were concerned, began anew in the fifth century. Since then many changes have taken place, but we can say with a fair amount of accuracy, that the Kentish people of to-day are descended from the Jutes, Angles, and Saxons who came over here in the fifth and sixth centuries.

The language now spoken in England is based on the speech of our Saxon ancestors. Of the Anglo-Saxon language there were two chief dialects, the Northern and the Southern. But after the Norman conquest, these dialects increased, so that we now recognise six dialects, of which Kent belongs to the Southern. The true dialect of Kent is found in East Kent, especially in the Weald, where many of the oldest forms of speech still survive. There we may hear many old English words that were used in Chaucer's days, and some of the peculiarities of speech are likely to continue for many years to come.

In connexion with this question of race and dialect, it is interesting to note that east of the Medway we speak of " A man of Kent," while one who dwells on the west of that river is called " A Kentish man." This distinction no doubt dates from the time when there were two kingdoms in Kent, one having its capital at Rochester, and the other at Canterbury.

The social life of the people of Kent was considerably changed by the settlement of skilled artisans in wool and silk from the Low Countries and France. The people from Holland and Belgium were known as Flemings and Walloons, and they settled at Cranbrook and in its neighbourhood, where they began the manufacture of woollen cloths. Broadcloth "halls" were built, which were the centres of great activity and helped to make Cranbrook one of the most prosperous towns in Kent. Cranbrook was at the height of its prosperity when Queen Elizabeth visited it in 1573.

The Walloons, Huguenots, and other refugees settled at Canterbury, Sandwich, and Dover at the time of the great religious persecutions in the Netherlands and France. In 1630 these visitors numbered 1300 persons, and after the Revocation of the Edict of Nantes, they were double that number. These strangers have left their mark

Church Street, Sandwich

on the social and industrial life of Kent, and although
the woollen and silk trades have passed away, the names
of French and Flemish descendants yet survive in many a
Kentish parish.

Now let us come to the present day, and consider the
character of the population of Kent. When the census
was taken in 1901, there were 1,348,841 people in the
ancient county of Kent; but as a portion of the county
is now included in London, the population of the present
administrative county was 961,139. How the population
has increased will be evident when we remember that in
1801 it was 258,973. The number of people to the
square mile in Kent is 618, against 558 for England and
Wales. The population of Kent has mainly increased in
the north-western, or Metropolitan portion, and in Dover,
Folkestone, Chatham, and Gillingham. Looking more
closely at the figures we find that about two-thirds of the
people in Kent live in towns or urban districts, and the
remainder in the villages or rural districts.

The census of 1901 shows that there were more
females than males in Kent. The former numbered
489,703 and the latter 471,436. People of foreign origin
living in Kent in 1901 numbered 4325, and these came
from all parts of the world. Where did these people
live when the census was taken? Of course the greater
number were in houses, of which 188,193 were inhabited.
There were many people, however, who were not living
in houses. For instance, about 18,000 were in military
barracks, and 9500 in naval barracks or on H.M. ships.
In workhouses, hospitals, asylums, and industrial schools

there were over 20,000; and in merchant vessels, inland barges, and boats there were 4000 people.

The census returns are of the most interesting character, and give everything of value that relates to the condition and occupation of the people of each county. This chapter has had many figures in it, but we will just finish by noting that there were 594 blind people, 545 deaf and dumb people, and about 8000 lunatics and imbeciles in Kent in 1901.

15. Agriculture — Main Cultivations, Woodlands, Stock.

We will now proceed to consider the agricultural products of Kent and their relative values. Every year the Board of Agriculture issue a report in which the vegetable products are arranged under the following divisions :—corn crops, green crops, clover, sanfoin and grasses for hay, grass not for hay, flax, hops, small fruit. The portion of land that does not produce any of these crops is said to be bare fallow, which accounts for 7554 acres in Kent. (See diagrams at end of volume.)

The corn crops are grown on 142,268 acres, and consist of wheat, barley, oats, rye, beans, and peas. Thus about one-seventh of the area of Kent is devoted to these crops, of which wheat, oats, and barley are the most important. There is very little rye grown in Kent, but beans and peas account for 19,000 acres.

The green crops consist of potatoes, turnips, swedes,

mangold, cabbage, rape, and vetches or tares, and they cover about 74,000 acres, or one-thirteenth of Kent. A small portion of the county, about one-thirtieth of its area, is devoted to the growing of clover, sanfoin and grasses. Some of this produce is for hay, and part is not for hay, the land being broken up in rotation.

The largest portion of agricultural land in Kent is given over to permanent pasture, or grass not broken up in rotation. This area of permanent pasture is no less than 427,645 acres, or nearly one-half of the whole county.

Very little flax is grown in Kent, but hops and fruit occupy a most important place. It is worth remembering that out of the 48,067 acres of hop plantations in England, no less than 30,655 are in Kent. As hop-growing is the most characteristic cultivation of the county, we will give a separate chapter to this subject.

Fruit is largely grown in Kent, and, owing to the proximity of London, a good price is obtained for the various kinds, such as cherries, strawberries, apples, pears, plums, currants, gooseberries, and filberts. Of these, the specially Kentish fruit is the cherry, which grows in the district on the border of the Thames in West Kent, and along the Darent and the Medway, but particularly at Gillingham, Teynham, and Polestead. There are few prettier sights than a Kentish cherry orchard in spring, when the trees are covered with their beautiful white blossoms. Cherries were certainly grown by the Romans while they occupied our country, but the best kinds of cherries were introduced into Kent from Flanders in

1540. It is said that the varieties grown in Kent are more numerous and much better than in any other part of England.

Market-gardens are numerous in the neighbourhood of London, and give employment to many who cultivate vegetables, asparagus, and tomatoes for the great markets.

The woodlands of Kent are both valuable and extensive, for they cover nearly 100,000 acres. Coppice woods, or those that are cut over at regular periods, account for three-quarters of the whole area of woodlands. West Kent abounds in woods and coppices, and some of those in the Weald are in their original forest state. The kinds of wood which grow in Kent are chiefly oak, beech, ash, chestnut, birch, and hazel. Immense quantities of hop-poles are obtained from some of the plantations, which also provide faggots for fuel, and piles for securing the sea-walls of the marshes.

Lastly, we come to the animals that are reared in Kent for various purposes. We will consider them in four divisions—as horses, cattle, sheep, and pigs. Of these, sheep are the most numerous, there being no less than 888,425 in 1905, or about 909 on every 1000 acres. This is a high proportion, and compares with 445 sheep per 1000 acres in England. There are many breeds of sheep in Kent, the Romney Marsh breed being both special and numerous.

Horses numbered 27,505, cattle 91,421, and pigs 65,177 in 1905. The greater number of the horses are used for agricultural purposes, while the cows are reared to supply milk to the towns of Kent and the metropolis.

16. Hops and Hop-picking.

We will devote this chapter to the hop plant, its cultivation, and its importance. From the previous chapter, it will be gathered that the Kentish hop-gardens occupy five-eighths of the area covered by all our English hop-gardens, and one-thirty-second of the whole area of Kent. It seems probable, however, that there will be a serious decrease in the area devoted to the cultivation of the plant, owing mainly to the large importation of cheap hops from America.

The hop plant was well known to the Romans, and was used by our old English forefathers in the making of their beer. Although it was native and early used in England, it does not appear that it was cultivated in our country till the early years of the reign of Henry VIII.

The hop grows luxuriantly, with an abundance of foliage. It has long, rough, twining stems and rough leaves. The part of the plant used in brewing, and sold as hops, is the ripe cone-shaped inflorescence of the female plant. The hop requires a deep, rich soil, and grows best in a sheltered position. The plants are obtained from cuttings taken from the old plants, and do not come to full bearing till the third year.

Hop-poles are set to the plants in spring, and removed at the end of the season in autumn. Formerly, hops were grown on straight poles, but the best are now grown so that after climbing its own pole to a height of about four feet, the bine is trained along strings stretched to

A Kentish Hop Garden

the top of the next row of poles. Thus, when the crop is ripe, the hop-garden looks like a number of greenhouses with roofs of foliage. As soon as the hops are ready to be picked, the strings are cut, and the upper part of the bine hooked down so as to be within easy reach of the picker. The cones are ready to be gathered when they have turned to an amber brown colour.

The hop-pickers have large baskets or bins, into which the hops are picked. Each of the baskets holds about 15 or 20 bushels, which is as much as the fastest hands can pick in one day. The contents of the baskets are emptied into sacks, and sent to the "oast" or drying house. These oast-houses are quite a feature in hopland. They are mostly built of brick, and, to a height of 14 or 15 feet, are circular. They terminate in a cone, above which is a cowled chimney, through which the vapour from the hops escapes.

The drying of the hops is a most important operation, and requires constant attention and much skill. In the lower part of the oast-house is a furnace, in which burns a clear fire of Welsh coal, or coke, or charcoal. Into this some sulphur is thrown, as this gives a better colour to the hops. Above the furnace are the drying-room and cooling-floor. On a circular floor, formed of strong wire-netting, and covered with coarse hair-cloth, the warm air ascends to the hop-flowers, which lie to a depth of two or three feet. When the hops begin to shrivel, they are taken off the kiln and laid on a wooden floor to cool. The drying never ceases during the time of picking, and is one of the most difficult parts of the preparation. On the cooling floor the hops are tightly packed into bags or "pockets," ready to be sent to the brewers.

Hop-gardens are to be found in most parts of Kent, but the best hops are grown at Barming, East Farleigh, and Hunton, in the neighbourhood of Maidstone. Between Faversham and Canterbury, and between Godstone and Ashford are, however, the chief Kentish hop-districts. There are many varieties of hops, such as the Goldings, Golden Hops, Grape Hops, and Farnham White Bine, but the Goldings are the best and richest.

In the season, a hop-garden is one of the most picturesque sights in Kent. Men, women, and children are all at work, from early morning to the close of day. The work is particularly healthy, and, if the "hopper" is diligent, as much as four or five shillings per day may be earned. At the end of August, crowds of people from the East and South of London, as well as from more distant parts, flock to Kent to take part in the hop-picking.

17. Industries and Manufactures.

It will be gathered from the previous chapters that Kent is mainly an agricultural county, and not in any sense a manufacturing county like Yorkshire or Lancashire. There are, however, some few important industries and manufactures carried on in Kent, and with these this chapter will deal.

Ship-building gives employment to a large number of men at Chatham, Rochester, Gillingham, and Sheerness.

5—2

Both Chatham and Sheerness are Royal Dockyards, and as such occupy a very important place in the history of our country. Chatham Dockyard has about 5000 men working in it, and these are employed in building and repairing our great war-ships, some of which cost £1,000,000. There are special store-houses, where giant masts, innumerable pulleys, cables, and anchors are kept. The whole dockyard covers upwards of 400 acres, and constitutes one of the wonders of our country.

Sheerness Dockyard, though not so large as Chatham, comprises wet and dry docks, store-houses, and other buildings. The harbour is generally full of vessels, and work is found for 2000 or 3000 men.

Woolwich Arsenal is a crowded hive of workers and employs from 12,000 to 14,000 men at a busy time. It is the only Government establishment of its kind in our country, so that it is of the utmost importance. The Arsenal covers 350 acres, and it has seven miles of road and as many of railway. A visit to this wonderful arsenal enables one to understand what it means to make the implements of war. In the Gun Factory men are making all kinds of cannon, from the lightest to those of 81 or 110 tons. The Carriage Department, with its 1000 men, is devoted to the construction of gun-carriages and all kinds of military vehicles. The Laboratory turns out shot and shell, rockets, fuses, and torpedoes ; and some idea of its work may be gained from the fact that 2,000,000 cartridges per week can be made in the building. The Stores Department houses war material of every kind, from which about 20,000 troops could be equipped at any time.   Woolwich Arsenal has 18 steam hammers, and turns out 6000 tons of guns in one year. The forty-ton hammer falls through a space of 15 feet and cost over £50,000.

It will thus be apparent that Chatham and Sheerness Dockyards and Woolwich Arsenal give employment to many thousands of skilled workmen.   At Erith, on the Thames, electrical apparatus of all kinds is made; and at Ashford the works of the South-Eastern and Chatham Railway employ a large number of men in making railway coaches and waggons.   Gunpowder is made at Dartford, Faversham, and Tunbridge; and lyddite, a very destructive explosive, takes its name from Lydd, a place used for experiments in firing.   Bricks are made at Sittingbourne, Milton, and Faversham, and employ not less than 5000 hands.   Stone is worked by quite a colony of masons at Maidstone, while cement is manufactured at Northfleet by 3000 men.

The manufacture of paper is one of the most important of Kentish industries.   With the single exception of Lancashire, Kent has more paper-mills than any other English county; and it also claims that the first mills for paper manufacture were established at Dartford by Sir John Spielman, early in the reign of Elizabeth.   The Kentish paper-mills are chiefly on the Darent and Medway, at Dartford, St Mary Cray, Maidstone, and Eynesford. There are nearly 5000 people in Kent who are engaged in the manufacture of paper, which is made from rags, esparto grass, and wood-pulp.   The special and better

Allington Castle, Maidstone

kinds of paper are made in Kent, and its manufacturers pride themselves on the best work.

Of late years there has been a great increase in jam-making. There are also breweries in the large towns, such as Dover, Margate, Greenwich, Dartford, Faversham and Maidstone.

Some of the Kentish industries that were once famous are now either decaying or dead. A little silk is still manufactured at Canterbury, but the cloth-works, that once existed at Cranbrook and neighbouring parishes, ceased towards the beginning of the nineteenth century. Some of the old cloth-houses have been converted into modern residences.

The iron-trade, too, was formerly carried on in the Weald, and in Elizabeth's reign there were furnaces at Cranbrook, Hawkhurst, Ashhurst, and Goudhurst. In some of the parishes in this district articles made of cast-iron from these furnaces may yet be found in some of the cottages and houses. The clothing trade and iron manufacture are now extinct, and other industries have taken their place.

18. Minerals—Exhausted Mining In= dustries.

Kent has no claim to rank as a mining county, but some very useful products are obtained that we may consider under the heading of minerals. Very good building-stone is quarried in several parts of Kent.

Maidstone

Quarries of "Kentish rag," a hard calcareous sand-stone, are largely worked near Maidstone. This stone is much used for churches and public buildings in London and the neighbourhood. A less durable stone is worked at Folkestone; while at Bethersden, five miles from Ashford, some good marble is obtained. From the North Downs hard grey chalk, used for work in the interior of churches, is dug.

Iron pyrites and septaria, or calcareous nodules, are collected at Folkestone and the Isle of Sheppey and used for various purposes. Chalk, gravel, and sand are dug in many places, and are employed for such purposes as building and road-making. The making of bricks and cement at Northfleet and Sittingbourne is referred to in the industries of Kent.

Much attention has been drawn to the finding of coal in Kent, owing to the work of some companies that have been formed during the last ten years. Boring close to Shakespeare cliff began in 1886, and coal measures were struck in 1890 at about the depth of 1100 feet. Since that year, other successful borings have been made at Dover, Waldershare, Fredville, and also at Ropersole, and seams of coal varying from 7 feet to 20 feet have been proved. In all these instances coal has been reached at a depth of about 1100 feet, and it is thus sufficiently near the surface to be worked. Professor Boyd Dawkins thinks there is every evidence of a South-Eastern coalfield, which will rank, when fully developed, among the important coalfields and cause centres of industry to be established in Kent. It is worth quoting the anticipations of Professor Boyd Dawkins, who writes thus : " It will probably attract a large population to the lonely downs, that will by their labour add to the wealth of the nation, and at the same time, convert the white into the ' black country,' or at all events into 'studies in black and white '."

There was a time in the history of Kent when it was the scene of iron-furnaces and iron-mills. In the reign of Elizabeth there were over 100 iron-furnaces in Sussex, Surrey, and Kent, and the timber in the Weald of Kent was cut down in such quantities that it was feared other industries would suffer. Towards the middle of the eighteenth century the iron industry decayed, owing to the difficulty of procuring fuel to work it successfully. In the neighbourhood of Cranbrook some of the farms bear such names as the Forge and the Furnace, and thus indicate the positions of some of the iron-works in the days of old. It is interesting to note that the fine railings of St Paul's Cathedral were made from iron from the mines at Lamberhurst, once a great seat of the iron manufacture of the Weald.

19. Fisheries and Fishing Stations.

The fisheries round the coasts of England are of great importance, and give employment to many thousands of people. Marine fishes are produced in enormous quantities without human aid, yet the amount of capital and labour required to capture them is very large. Sea-going vessels and boats are costly, and expensive machinery has to be carried on board. The vessels and gear are subjected to very hard wear ; and sometimes both vessels and gear are lost altogether.

The capture of fish and its consumption have greatly increased since the introduction of steam. Before the age of steam very little of the fish found its way beyond the coast towns, where it was sold by the fisher-folk from house to house. Now the fish is no sooner landed than it is packed, and carried by the railways to all parts of our country. Of course, the increased demand for fish is also due to the rapid growth of our population, who are glad to buy a cheap and palatable food.

Although there are fisheries round the south, east, and west coasts, it is worth noting that those on the east coast are four times more productive than those on the west. This is chiefly owing to the great shoals of the North Sea, from which we obtain so large a supply of fish. The fisheries on the west coast are at a disadvantage because they are so far from the large and populous centres.

As we should expect from their situation, the fisheries of Kent, both in the estuary of the Thames and in the North Sea and English Channel, are of considerable importance. Many of the Kentish coast-towns have more or less interest in the sea-fisheries, and the number of men employed in the season is very large.

We may divide the sea-fisheries of Kent into four classes. First there are the North Sea, or deep sea trawlers, and boats from Ramsgate, Dover and Folkestone are chiefly engaged in this, the most important branch. Secondly, there is off-shore and in-shore fishing, which is practised by the fishermen of Margate, Deal, and Dungeness. Thirdly, there is estuarine fishing in the widest part of the Thames estuary, Sheerness and

Old Houses on Deal Beach

Queenborough being the ports chiefly concerned with this class of fishing. Fourthly, there is the shell-fish branch, in which Whitstable and Faversham are interested.

The methods of catching the fish vary with each class. For the deep sea fishing the trawl-net and drift-net are used, although on the Dogger Bank some fish such as cod are caught singly on long lines and not with nets. When the fish were caught, the old method was to place them in the "well" of the boat, where they were kept alive by a constant change of water. Since the use of steam-carriers and ice, these well-boats are not of so much importance, although they still play their part on the Kentish coasts.

The other methods of catching fish round the Kentish coast are by means of shrimp-nets, dredge-nets, kettle-nets, and crab and lobster pots. Kettle fishing is employed chiefly for the capture of the various species of flat-fish which frequent the shallow waters covering the sands at high tide. Kettle-nets are about 120 yards long and 4 feet high. They are fixed in position by stakes driven into the ground, and to these the head and ground-ropes are attached. The nets are in the form of the letter V, with the apex of the V, which has a purse, pointing away from the shore. As the fish follow the rising tide they are carried between the nets, and arrive at the hollow of the V. On their return on the falling tide, they are carried into the purse at the apex. The nets are visited when the tide is falling, and the fish are quickly removed.

The methods of catching crabs and lobsters are well known to all visitors to our sea-side towns. Crabs and lobsters are taken in traps, which are dome-shaped cages made of wicker-work, or netting stretched on a strong frame. The trap is baited with pieces of fish, and there are openings in the sides, in the form of funnels, projecting into the interior of the trap. The oyster-dredge is like a small trawl, but the mouth is made by a rectangle of iron bands, and the net is usually composed of iron rings linked together.

It would not be possible to name all the varieties of fish caught round the Kentish coast, so we will mention only the principal. Flounder, plaice, dab, sole, halibut, turbot, brill, cod, haddock, whiting, ling, herring, sprat, whitebait, shad, and pilchard are among the best known. Among shell-fish may be mentioned oysters, scallops, mussels, cockles, periwinkles, and whelks; shrimps and prawns are most abundant; and, in the Thames, smelt and mullet are caught.

The Thames fisheries are not of the value they once were. The Thames has been poisoned by sewage, which has destroyed the only salmon-river in the county, although some individuals occasionally attempt to pass up stream. The tiny fish known as whitebait is caught in the Thames and sent to London, where it has a ready sale. The Ministerial whitebait dinner at Greenwich was once a very important function, but has now fallen into abeyance.

Some of the largest English oyster-beds lie off Whitstable, and the "natives" from that town always command a high price. The "spat" or young brood is frequently brought from a great distance and laid in the bed, where they remain for three years before they are brought to market.

Such large fish as the sturgeon and shark are occasionally found in Kentish waters. The sturgeon is sometimes captured in the Thames, and is then presented to the Lord Mayor of London. In recent years more than one whale has come south to the Thames. In August, 1898, the dead body of a sperm whale 42½ feet long was washed ashore at Birchington ; and in 1859 a rorqual was found at Hope Reach, in the Thames, near Gravesend, and another was captured in the Medway in 1888.

20. Shipping and Trade—The Chief Ports. Extinct Ports.

The chief sea-ports of Kent are Gravesend, Chatham, Sheerness, Ramsgate, Deal, Dover, and Folkestone. Gravesend has been well named the "Water-gate" of London, for it is the entrance to the Port of London. All ships bound for foreign ports must call at Gravesend to take a pilot on board. There are 359 Trinity pilots, of whom many live at this port. Inward-bound ships are boarded by Custom House officers ; and emigrant ships anchor here and undergo inspection.

Chatham is one of the "Three Towns," as Strood, Rochester, and Chatham are called. It is naturally of great importance as one of our dockyards, and because of its position on the Medway. Sheerness is also a Royal Dockyard, and commands the entrance of the Medway and the Thames.

Ramsgate is really the harbour of refuge for the Downs. Its harbour of more than 50 acres is enclosed by two broad stone piers ; and there are occasions when it presents a very animated appearance. Here may be seen private yachts and commercial shipping, and as many as 400 vessels have been accommodated in this harbour.

Deal is a little port of some interest from its position with regard to the Downs and the Goodwin Sands. It is one of Lloyd's stations, and has a time-ball tower, by which ships in the Downs may correct their chronometers.

In every way, Dover is the most important of the Kentish sea-ports. This is mainly owing to its position with regard to the Continent. It has not a good natural harbour, but large sums have been spent to improve its accommodation. Its trade chiefly depends on the continental service, and it is the terminus of the South-Eastern and Chatham Railway. Dover has ship-building, rope and sail-making, fisheries and some coast traffic. Eggs and all kinds of agricultural produce are largely imported from France. The harbour has been much improved in recent years, and consists of two docks and a tidal basin. The continental mail-boats depart from the fine Admiralty Pier, on to which the boat trains run. This pier, 600 feet long, forms one side of a national harbour of refuge, which is now being constructed. This harbour, with a water area of over 600 acres, will take the largest vessels, and is estimated to cost £3,750,000. The Dover Harbour Board are also constructing a graving dock, a commercial harbour, and a water station where continental passengers will be able to join their trains under cover. Some idea of the importance of Dover as a sea-port may be gained from the fact that its yearly tonnage, entered and cleared, amounts to nearly one million tons.

Folkestone is a thriving sea-port, and has much trade with the Continent. In one year no less than 750,000

Dover Castle

B. K.

6

tons were entered and cleared. Great improvements have been made in the harbour by the South-Eastern and Chatham Railway Company, who are developing both passenger and goods traffic.

Kent is one of those counties that have seen the rise and decay of several sea-ports. The ports already

Sarre Wall

mentioned have shown signs of steady progress ; but there are some that have really ceased to be sea-ports, or are such only in name. From Reculver on the Thames to Richborough on the east coast, there was once a sea-channel known as the Wantsum, filled by the water of the Stour. This sea passage was a short cut for vessels bound for Sandwich and London. The Wantsum was closed at the end of the fifteenth century, and then Sandwich, Richborough, and Reculver lost their importance as sea-ports.

Sandwich was once the resort of vessels of all sizes from many quarters. The action of the tide silted up the haven, and no attempt was made to save it. Now it is nearly two miles from the sea, or four miles by the winding course of the Stour. Its ancient glory has departed, and visitors have some difficulty in believing that the present quaint town was once a port of great reputation.

Hythe is another port from which the sea has retreated. It is now a clean, well-ordered town, with a bank of shingle one mile in width before the sea is reached.

Romney is also an extinct port, for it stands more than one mile from the sea. Sturry, Fordwich, Stonar, and Sarre have all lost their positions as ports, owing to the vast changes that have been worked by the sea in this part of Kent.

21. Cinque Ports. Trade Routes.

A special reference must be made to the Cinque Ports, some of which are in Kent. Originally there were five ports, hence the French name, but afterwards two others were added to the confederation. The Cinque Ports are Hastings, Sandwich, Romney, Hythe, and Dover, and the two Ancient Towns are Winchelsea and Rye. Besides these seven ports, almost every town from Pevensey in Sussex to Faversham in Kent was attached

6—2

to the Cinque Ports, and we find that these lesser towns were called " limbs." Some of the " limbs," such as Tenterden, were far from the sea.

The origin of the Cinque Ports is not quite clear, but most writers are agreed that they were fortresses that the Count of the Saxon Shore had under his control to guard the landing places round the south-eastern coast of England. Whatever their origin, they have filled an important place in our history, although Dover is at present the only one of them in a flourishing condition.

The Cinque Ports had in early days to provide so many ships and men to serve the King for a certain period in each year, and in return for these duties they had many privileges. They were all self-governed, and their freemen were allowed to trade free of toll in all English boroughs. The men were exempt from military duty, and all offences were tried before the Lord Warden.

In the fifteenth century the Cinque Ports began to decline, for the sea had retreated from the towns, and King Henry VII formed a new Royal Navy, so that the assistance of Cinque Port seamen was no longer needed. The Cinque Ports, however, still keep some of their old privileges and retain many quaint customs. Down to quite recent times Walmer Castle was the residence of the Lord Warden, but has now been handed over for public use. The present Lord Warden is Lord Brassey, who fills an office that has been held by such great men as Pitt, Wellington, Lord Palmerston, and Lord Salisbury.

It will thus be seen that in the past Sandwich and Dover were the chief ports with regard to the Continent.

Sandwich has declined, but Dover has so improved its position, that it is now the premier port of Kent. It is the chief trade-route from Kent to the Continent, and fine vessels leave here every day for Ostend and Calais. The route from Dover to Calais is the shortest, and the 21 miles is covered in favourable weather in less than one hour, while the whole journey from London to Paris takes under 7 hours. Another popular route to the Continent is from Folkestone to Boulogne, a distance of 26 miles.

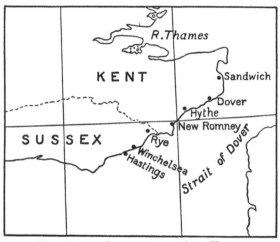

The Cinque Ports and Ancient Towns

Port Victoria and Queenborough in North Kent are the headquarters of the Royal Mail Route to Holland, Germany, and North Europe. The ships from these two ports trade with Flushing, and the sea journey is 205 miles.

22. History of Kent.

The county of Kent has been the scene of some of the most important events in our history. Its position at the narrowest part of the Channel brought its people from the earliest times into close connexion with those on the Continent, and it thus became the scene of three important landings, each of which stands out as a landmark in England's story.

In 55 B.C., Julius Caesar landed in Kent, probably between Walmer and Deal, and at once began the Roman invasion of Britain. He was stoutly opposed by the Britons, and was forced to leave England for a time. Returning in 54 B.C., Caesar was more successful, but he soon left our shores, never to return. The Romans again invaded Britain under Claudius in 43 A.D., and remained from that date to 410 A.D. as rulers of our land.

The second landing in Kent was that of our English forefathers in 449 A.D. It is generally believed that the first tribe to land in Kent were the people known as the Jutes, under the leadership of Hengest and Horsa. The site of their landing is supposed to have been at Ebbsfleet in Thanet. The English conquered our land, which was divided into various kingdoms. The Jutes allowed Kent to retain its Keltic name, and it was also made one of the kingdoms. Perhaps, at one time, it consisted of two kingdoms—East Kent and West Kent —which were gradually united under one ruler. There were 13 Kings of Kent, of whom two were Bretwaldas.

Baldred was the last King of Kent, and he was driven from his throne by Egbert, King of the West Saxons, who seems to have become the first King of England.

The third landing in Kent was that of Augustine and his followers at Ebbsfleet in 597 A.D. Augustine converted Ethelbert the King of Kent and large numbers of the English, so that he was the means of re-introducing Christianity into England.

Cross at Ebbsfleet, erected to commemorate the
landing of St Augustine

When the English had settled in our land, the Northmen from Norway and Denmark descended upon our coasts, and effected a landing in Sheppey in 832 A.D.

After a few years they invaded Thanet and besieged Rochester. In 1011 the Danes took Canterbury, plundered the city, burned the Cathedral, and put the people to the sword. A year later they cruelly murdered Archbishop Ælfheah (Alphege) at Greenwich.

Godwine, Earl of Wessex, who had large possessions in Kent, married Edith, Edward the Confessor's daughter. It was during Godwine's earldom that the men of Dover rose against Eustace, and the Earl of Wessex, refusing to punish the rioters, was forced to leave England.

When William the Conqueror fought against Harold at Hastings, the men of Kent were put in the van of the battle, and seem to have suffered terrible loss. After the battle, William visited Romney, Dover, and Canterbury on his way to London. The people of Kent did not please him, and, as a result, he gave nearly all the land of the English owners in Kent to his own followers. Odo, Bishop of Bayeux, was the first Norman Earl, and his possessions in Kent were most extensive.

The next great event in the history of Kent after the Conquest was the murder of Archbishop Becket in his own cathedral at Canterbury on December 29, 1170. The shrine of Becket drew pilgrims from far and near, and from every Kentish port people landed and found their way through the wild forest land of Kent to do homage and offer gifts to the " holy blissful martyr."

The end of the fourteenth century witnessed the outbreak of Wat Tyler's rebellion. The rising began at Dartford, where Wat Tyler murdered a poll-tax collector. It is said that no less than 100,000 rebels

TERRA ẼPI BAIOCENSIS.

In lest de Sudtone. In Achestan hund.

De ẽpo Baiocensi ten̄ hugo de porth hagelei.
p dimidio solin se defd. t̄ra.ē. In d̄nio sunt.II.
car̄.7 XIIII. uilli cū.IIII. bord h̄nc.IIII. car̄. Ibi.III. serui.7 XII.
ac p̄a.7 uū molđ de.XX. sol.7 una dena silue de.V. porc.
Totū m̄ ual̄ m̄.XV.lib̄. de.XX. in ora.

In hac m̄ ten̄ uū hō.XX. ac̄ t̄re ualentes p annū.V. sot. Vluret
uocat̄. nec ptin̄ ad illud m̄. neꝗ poruit habere d̄nm p̄ rege.

In elto ten̄ Dunescamp de ẽpo. p X. solins se defd.
t̄ra.ē.XIIII. car̄. In d̄nio sunt. III.7 XXX III. uilli cū.IIII. bord
h̄nc XIII. car̄. Ibi uū miles 7 X. serui.7 XL. ac̄ p̄a. Silua
III. porc.7 V. piscarie de.XXX. den̄.7 m̄. que seruit ad halla.
7 una heda de. V. solđ.7 IIII. den̄. De silua hui m̄ tene
Bucard in sua leuua q̄d ual̄.IIII. solđ.
Totū m̄ ual̄ XX. lib̄. 7 m̄ ual̄ XXX V. lib̄.

Radulf fili Turaldi ten̄ de ẽpo Eaclei. p uno solin se
defd. t̄ra.ē In d̄nio sunt.II. car̄. 7 IX. uilli cū.VI. ·]
cot h̄nc.III. car̄. Ibi.III. serui.7 silua.X. porc.
Totū m̄ ual̄.III. lib̄. 7 m̄.C. sot. ꝗdā mulier tenuit.

Radulf ten̄ de ẽpo Eddintone. p dimid solin.
t̄ra.ē.I. car̄.7 ibi.ē cū.IIII. bord 7.II. serui.7 ibi.I. molin
de.XXVIII.sot. Totū m̄ appciat̄.IIII. lib̄. T.R.E. paru ual̄.
Lestan tenuit de rege E. 7 post morte q̄ uertit se ad
aliud. 7 m̄ est in calupnia.

Facsimile of Domesday Book

followed Wat Tyler in his march to London over
Blackheath, in June 1381. Wat Tyler was killed, and
many of his followers were pursued into Kent, and either
died in fighting the King's soldiers, or were executed.

The middle of the fifteenth century saw a second
rising in Kent. This time the leader was Jack Cade,
a tanner of Ashford. With a following of 30,000 men,
he defeated the Royal army at Sevenoaks, and then
marched on London. There the rebels did considerable
damage, but eventually were overcome. Jack Cade fled
to Sussex, where he was slain in 1450.

Another rebellion in which the men of Kent were
concerned took place in 1554. The leader of this
insurrection was Sir Thomas Wyatt, a gallant young
knight, of Allington Castle, near Maidstone. The cause
of this rebellion seems to have been the marriage of Mary
with Philip of Spain. Wyatt raised 10,000 men whom
he gathered at Maidstone, and led to London. Here
they were routed and dispersed, and Wyatt, after a
desperate struggle, was taken prisoner at Temple Bar.
He was kept in confinement for some time, and then
executed.

The period of our history which goes by the name of
the Reformation was a time of great change in Kent.
Fisher, Bishop of Rochester, was executed in 1535, and
from that year onwards the new religious movement had
free play. Henry VIII gave his chief agent, Thomas
Cromwell, unlimited power; and this "Hammer of the
Monks," as he was called, was the means of closing all
the religious houses in Kent. Their annual income

was said to amount to £9000, and no doubt much of this sum went into the coffers of the King and his friends.

The accession of Queen Mary stopped the persecution of the Roman Catholics, only to bring like troubles on the Protestants. Many Kentish people were burned at Canterbury and Dartford, and there is a monument to their memory in the former city.

The year 1588 witnessed the attempted invasion of England by the Spanish Armada. The men of Kent did their duty right well in helping to defeat the hated Spaniards. The Cinque Ports gave ships and money ; beacons blazed on every hill :

" And eastward straight from wild Blackheath the warlike errand went,
And roused in many an ancient hall the gallant squires of Kent."

We need not dwell on the great naval fight, nor on the bravery of our seamen during those eventful days at the end of the glorious month of July, 1588. That great victory is one of the most famous in the annals of our land.

During the civil war between Charles I and the Parliament, Kent was mainly royalist. The chief event in Kent was the siege of Maidstone by Fairfax in 1648. The capital of Kent offered a most stubborn resistance before it was forced to surrender to the Ironsides.

The reign of Charles II was disgraced by the presence of the Dutch fleet in the Medway. The English ships did not attempt a proper resistance, and the only wonder is that the Dutch did not do more damage. After some skirmishes De Ruyter sailed away, but a feeling of shame was in the hearts of all true Englishmen that it was possible for the honour of England to be thus lowered.

It is not possible in this chapter to give an account of many other minor events in the history of Kent. The "Royal County" has made wonderful progress, both in wealth and population, and it will ever be the wish of all true "Men of Kent" and "Kentishmen" that it may continue to flourish.

23. Antiquities — Prehistoric, Roman, Saxon.

The earliest history of the people who dwelt in Kent is not derived from written records, but from the relics or antiquities that have been dug up in various parts of the county. The earliest written records of Kent and its people do not carry us back more than 2500 years ago, so that prior to that period, and even after it for some time, we are dependent for our knowledge of the dwellers in Kent on the traces they have left of their handiwork.

Antiquaries have divided the earliest portions of our country's history into the Stone Age, the Bronze Age, and the Iron Age. After this last period we generally speak of Roman antiquities and Saxon antiquities, which correspond with the history of our land from 55 B.C. to 1066 A.D. Antiquities representing all these five divisions of time have been found in Kent, but the Roman antiquities are among the most interesting relics in England.

Palaeolithic Flint Implement
(From Kent's Cavern)

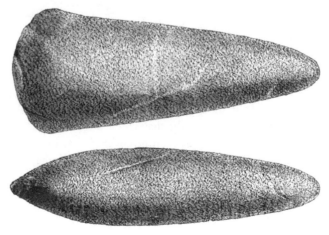

Neolithic Celt of Greenstone
(From Bridlington, Yorks.)

Kent has yielded some good specimens of the work of the Stone Age, such as arrow and spear-heads of flint, and axes and hammers of stone. These have been found in various parts, but especially in the valley of the Thames, at Swanscombe, Greenhithe, and between Reculver and Herne Bay. The Bronze Age marks the period when early people began to use metal, and specimens of bronze implements, such as celts and swords, have been found at

Ancient Cromlech : Kit's Coty House

Woolwich, Dover, Canterbury, Ashford, Hythe, and elsewhere. It is difficult to say when the Bronze Age ended, but there is no doubt that when once a metal such as bronze was used, it was soon discovered that iron could be fashioned into weapons with sharp edges. Perhaps the use of bronze was continued for ornaments and trinkets for some time into the Iron Age, which would be the period when the Britons were living in Kent.

One of the most remarkable relics of these early ages is a dolmen or cromlech, known as Kit's Coty House, which stands on the summit of a hill, not far from Aylesford. This is supposed to mark the site of an ancient burying ground, for the surrounding hills are covered with smaller cromlechs. The stones of this cromlech are of vast size, and one of them is said to weigh 11 tons.

Among the British remains found in Kent are coins and pottery, which have been obtained from barrows or mounds in various parts of the county. Camps or earthworks of the Britons have been traced, and the site of a British village at Ramsgate has been laid open. The Roman road in Kent ran from east to west; and on either side of this road, which is called Watling Street, some of the most interesting Roman relics have been found. It was in this district that the Romans built their villas and placed their cemeteries, and as a consequence the plough and the spade are constantly turning up fine relics of this period.

Pottery of the Roman period has been found in great quantities at Upchurch on the Thames, and at Dymchurch on Romney Marsh. Both these districts were the sites of extensive potteries, where such articles as urns, vases, drinking cups, and household utensils were made. The Romans also manufactured glass, and glass cups and bowls have been found at Richborough and Hartlip. The Roman coins were probably made at a mint at Richborough, for hoards of coins, gold, silver, and brass, have been found in the neighbourhood of that camp.

The Romans had three strong fortresses in Kent, and the remains of them remind us of the power of the conquerors of the Britons. The massive walls of Richborough, Lymne, and Reculver are among the most remarkable Roman remains in Britain. The Pharos, or lighthouse, is another Roman monument that still stands at Dover. It was probably the centre of an entrenchment, and formed a beacon to guide the Roman ships into the harbour.

Richborough Castle

The remains of the Saxons in Kent are both numerous and interesting. Saxon graves and cemeteries have been opened at Osengal, Sarre, Darenth, Faversham, Minster, and elsewhere, and in them have been found spears, shields, knives, swords, and ornaments, such as beads, brooches, etc. In other parts peculiar vessels made of wood, metal, and glass have been unearthed, and may now be seen in the museums at Maidstone and Canterbury. The earliest Saxon coins were called sceattae. They were of silver, and have been found at Richborough, Reculver, and in other parts of East Kent.

It was during the Saxon period that churches and monasteries were first built in Kent, but of these antiquities we shall read something in the section on Architecture.

24. Architecture — (*a*) Ecclesiastical— Cathedrals, Churches, Abbeys.

We will consider the architecture of the buildings in Kent under three divisions, viz. : (*a*) Ecclesiastical, or buildings relating to the Church ; (*b*) Military, or Castles ; (*c*) Domestic, or houses and cottages.

There is one fact worth noting with regard to all these classes of buildings, and that is that—as indeed is universally the rule—the architecture of the county has been affected by the materials accessible. Thus we find that stone, wood, flints, and bricks are used either because they could be easily obtained, or because of the wealth or otherwise of the builders.

Now with regard to the ecclesiastical buildings, let us consider first the churches and cathedrals, and then glance at the remains of the abbeys, monasteries, and other

religious houses.   The churches of Kent are of various
styles and of different ages, so that it will be well to
classify them as Saxon, Norman, Early English, Decorated,
and Perpendicular.

Patrixbourne Church

Towards the end of the twelfth century the round
arches and heavy columns of Norman work began gradually
to give place to the pointed arch and lighter style of the
first period of Gothic architecture which we know as
Early English, conspicuous for its long narrow windows,
and leading in its turn by a transitional period into the
highest development of Gothic—the Decorated period.
This, in England, prevailed throughout the greater part of
the fourteenth century, and was particularly characterised
by its window tracery.   The Perpendicular, which, as its
name implies, is remarkable for the perpendicular arrange-

ment of the tracery, and also for the flattened arches and the square arrangement of the mouldings over them, was the last of the Gothic styles. It developed gradually from the Decorated towards the end of the fourteenth century and was in use till about the middle of the sixteenth century.

Kent is interesting in its many churches, and it has good specimens of every style. Speaking generally, it may be said that the external work is very plain, and that the rich ornamental work is found within the church.

The materials used in the construction of the churches vary with the locality. Thus in the Weald, and in part of West Kent, Kentish rag and Bethersden stone are largely used, though sometimes intermixed with flints. In parts of East Kent rough flints and chalk are the usual material, and the churches are generally smaller and more rude than in other parts of the county.

One of the most noticeable features in Kentish churches is the square tower of stone with a turret at one angle. In the larger churches there is usually a large aisle on each side of the chancel, sometimes wider than the nave. The aisles have often separate roofs, which are covered with tiles.

The smaller churches have often no aisles ; many have no steeple, but only a wooden belfry, or else a low tower of flints, surmounted by a wooden spire. Kent has the unusual distinction of possessing two cathedrals. Canterbury, one of the most magnificent in England, can hardly be said to have any features peculiar to Kent, for it is in every sense a national cathedral. Rochester, though not of the first rank, has very much of Kentish character, inasmuch as its external work is of Kentish rag with flints.

As we might expect, there are very few Saxon churches in Kent. The probability is that the first churches were of wood, and thus perished through the fire of the Danes or by decay. Saxon churches may be seen at Lyminge, Swanscombe, and Cheriton, and part of the church in Dover Castle is of this period.

Barfrestone Church

There are many examples of Norman work, both plain and rich. Among the best Norman churches are St Margaret-at-Cliffe, Minster in Thanet, the nave of Rochester Cathedral, Barfrestone, Paddlesworth, Patrixbourne, and Darenth. These are all interesting, but Barfrestone is probably the finest.

When we come to the Early English style, the examples are very numerous. The magnificent choir of Canterbury Cathedral is a very fine example, and the choir and transept of Rochester Cathedral both belong to this style. Other early English churches are at Hythe, St Martin's in Canterbury, and St Clement's at Sandwich.

The Decorated style prevailed in the reigns of the three Edwards, from about 1300–1399. The churches in this style that deserve most attention are Hever, Barham, Herne, Stone, and Chilham.

The Perpendicular style was common from the reign of Richard II to 1547, or thereabouts. The churches in this style are generally not rich, but the most highly finished specimen is the western portion of Canterbury Cathedral. The Kent churches of this period have large porches, generally of wood, and there is a good deal of wood screen-work. The churches at Ashford, Chislehurst, Cranbrook, and the nave of Canterbury Cathedral are excellent specimens of Perpendicular work.

Before the Reformation religious houses such as abbeys, priories, nunneries, and hospitals were numerous in Kent. Henry VIII determined to close these religious houses, and employed Thomas Cromwell as his agent in suppressing the monasteries. The smaller ones were first closed, and then the great abbeys shared the same fate. Some of these buildings were fine specimens of the skill of the architect, but to-day we have only gateways or towers, or some other fragments to remind us of their former beauty.

Among the monastic buildings of which portions

remain, we may mention St Augustine's Abbey and the Priory of Christ Church at Canterbury, St. Martin's

Gateway to St Augustine's Monastery, Canterbury

Priory at Dover, Malling Abbey, and St Radegund's Abbey near Dover.

25. Architecture—(b) Military—Castles.

There were castles in England before the Norman Conquest, but by far the larger number of English castles were built during the Norman period. Indeed, it is said that as many as 1100 were built in various parts of the country at that time.

These Norman castles were not all of the same size and importance. Some were royal castles, built for the defence of the country and placed under the charge of a constable or guardian. Others were built by the Norman barons for the defence of their own possessions, and became at once the terror of the neighbourhood.

Before we deal with the Kentish castles in particular, let us consider some general details connected with these buildings. A castle of the best construction consisted of a lofty and very thick wall, with towers and bastions, enclosing several acres, and protected by a moat or ditch. Within this area were three principal divisions. First, there was the outer bailey, or courtyard, the approach to which was guarded by a towered gateway, with a drawbridge and portcullis. In this bailey were the stables, and a mount of command and of execution.

Second, there was the inner bailey, or quadrangle, also defended by gateway and towers. Within this second court stood the keep, the chapel, and the barracks. Third, there was the donjon or keep, which was the real citadel, and always provided with a well.

With regard to the position of the Norman castles, we find that in Kent many are built on low ground and surrounded by waters, collected by damming up some stream. In a country where there were no inaccessible

Tonbridge Castle

rocks, the builders had to be content with the securities furnished by water and the low ground. The castles at Leeds and Tonbridge are both good illustrations of these methods of defence.

Rochester Castle

Dover Castle has a fine position, for it is built on the high cliffs overlooking the sea. The area covered by this castle is 35 acres, and although it has been much altered, it is a good type of a Norman castle, with keep, courts, watch-towers, and underground passages.

Rochester Castle has a Norman keep 70 feet square and 104 feet high. Like Dover, this castle is noteworthy for the strength and massive character of the masonry. Canterbury Castle has the third largest keep in England, and is said to have been built by William the Conqueror.

Leeds Castle

Saltwood Castle, near Hythe, is most picturesquely situated, with low wooded hills on either side and the sea in front. There was a broad, deep moat within the outer walls; and beyond the moat, which is now dry, is the gate-house, flanked by two circular towers. Hever

Castle was built in the fourteenth century, and has been recently restored.

King Henry VIII built several castles on the Kentish coast for the protection of our land. Their sites were at Sandown, Deal, Walmer, and Sandgate, and they were low structures, with rounded walls, and masonry of great strength. Upnor Castle, built by Queen Elizabeth for the defence of the Medway, is now only a store-house. It was here that two battles were fought with the Dutch in 1667, and several English ships were destroyed.

26. Architecture—(c) Domestic.— Famous Seats, Manor Houses, Cottages.

The necessity for castles, or fortified houses, passed away after the Wars of the Roses, and we find that in Tudor times the great houses of the nobles were built less like fortresses and more as comfortable homes for the owner, his family, and his servants. Many of these fine Tudor mansions are still standing in various parts of England, and bear testimony to the skill of their architects and the good workmanship of the builders.

The style of these lordly houses was very different from that which had till then prevailed, for Italian ideas were coming into favour. The general plan was to build a big house round a quadrangle. The hall was in the middle, and on either side were the wings. The materials used largely depended on the wealth and taste of the builder, or on the locality. In Tudor times, and still more in Stuart times, brick was much used, as well as wood and stone.

Before dealing with the mansions of Kent, let us glance for a while at the ancient palace of Eltham. It is interesting as having been the residence of some of our English Kings; and from the time of Henry III to that of Henry VIII it was the scene of royal splendour and feasting, and the meeting-place of several parliaments and councils. The chief remaining part of the palace is the fine hall with a magnificent roof of oak. The moat also remains, and this is spanned by an ivy-covered bridge.

Kent has many great and famous mansions, and also a number of picturesque manor houses and farm houses. It was considered a very desirable county, as it was near the metropolis, and was fruitful, well-wooded, and very pleasant. Hence we find that the fifteenth and sixteenth centuries witnessed the building of the best of these stately mansions, some of which remain almost unaltered to this day.

One of the oldest and most charming of the old houses of Kent is Ightham Mote, or Mote House, 10 miles west of Maidstone. It is built round a square courtyard, and is surrounded by a deep moat. The building is of various styles, ranging from the Perpendicular of the reign of Edward III to the Tudor of the sixteenth century.

Knole House, near Sevenoaks, is perhaps the largest of the Kentish mansions. The earliest parts were built in the fifteenth century, but the main portion is chiefly Elizabethan. It is a quadrangular house, with embattled

Ightham Mote

gateways and square towers, and is most beautifully
situated in an extensive park, which has deep hollows
and a splendid variety of trees.

Penshurst Place near Tonbridge, and Cobham Hall
in the north of Kent, are two of Kent's finest mansions.
The former is specially famous as the home of the
Sidneys, and the latter as being the work of Inigo Jones,
the great architect.

The numerous farm houses of Kent that date from
the 16th or 17th century were in many cases constructed
of timber.   Perhaps, however, the most picturesque
domestic buildings left in the villages are the old
fashioned cottages, which please the eye with their quaint
gables and great chimney stacks.   They present a vivid
contrast to the ugly and monotonous buildings of the
present age that are raised in our towns, both small and
large.

It would be difficult to find a county that can boast of
so many charming villages, with their picturesque cottages,
as Kent.   Next to the village church it is probable that
the country cottage appeals most to the heart and
imagination of an Englishman, and in every part of Kent,
in the village and on the outskirts of the town, there still
remain these witnesses to the pride of the village mason
and carpenter.

The materials used for cottage building varied with
the locality, but in Kent we find that brick and timber,
plaster, weather-boarding, and weather-tiling were used
for the walls, and tiles for the roofs.   In the east of
Kent cottages built of flint and stone, roofed with tile,

Doorway, Cobham Hall

are common. The beauty of Kentish cottages lies in the individuality shown in their construction, and the effective use of the materials common to the district.

The most beautiful examples of cottage architecture in Kent may be seen at Cranbrook, Goudhurst, Penshurst, and Chiddingstone. In the towns there are some good examples of picturesque architecture, and a walk through Canterbury, or Sandwich, or Deal, will show how much pride our forefathers took in the sound work and good appearance of their houses.

The great difference between these ancient cottages and houses and those of modern date is that the former harmonise with their surroundings, while the latter do not. These old cottages gain in beauty with age, but the ugly modern construction of yellow bricks and slate roofs will certainly not improve as time advances. No painter would put on canvas one of the mean houses of our new towns, whereas the old Kentish cottages are often sought after for their artistic effects.

27. Communications — Past and Present—Roads, Railways, Canals.

Before we deal with the present condition of the roads and railways of Kent, it will be well to give a glance at the past history of the internal communications. Probably the most ancient road through Kent is that known as "The Pilgrims' Way." This is generally considered to be a British road and runs from

Old Houses, Chiddingstone

B. K.

8

Southampton to Canterbury. Traces of it are yet visible throughout Kent, Surrey, and Hampshire, and its course is often marked by long lines of Kentish yews, usually creeping half-way up the hills and avoiding for the most part the towns and villages, and the regular roads. It is called "The Pilgrims' Way," as it is the route that was followed by pilgrims to Canterbury.

It is well known that the Romans were great in making roads. Kent was, perhaps, the first county that gave them experience in England, and we find the roads they formed have been continued to the present time. The Roman roads were thoroughly made and were of various kinds, as military roads, branch roads, and private roads, but there is no doubt that the roads were constructed in the first instance for the convenience of the soldiers.

The old Roman road leading from Lymne to Canterbury is still called Stone Street. Two other roads ran from Richborough and Dover to Canterbury, where they united. The road to London, or Watling Street as it was called, then followed its course by Faversham, Sittingbourne, Rochester, and Gravesend. It is thus evident that the Romans were intent, first upon uniting their military towns in Kent, viz. Richborough, Dover, Lymne and Reculver, and then pushing a good road, as straight as possible, through the great Kentish forest to London.

We can now come down to more modern times, when we shall find that the roads of Kent have been much improved, both in character and extent. It is, however, only in the nineteenth century that a real improvement was made. In the eighteenth century it was considered quite an enterprise to journey by coach from London to Dover. Here is an advertisement of a time-bill which shows how such a journey was accomplished:—

"London Evening Post. March 28, 1751.

A Stage Coach

will set out

For Dover every Wednesday and Friday from Christopher Shaw's, the Golden Cross, at four in the morning to go over Westminster Bridge to Rochester to dinner, to Canterbury at night, and to Dover the next morning early: will take up passengers for Rochester, Sittingbourne, Ospringe, and Canterbury—and returns on Tuesdays and Thursdays."

When we compare this leisurely mode of travelling with our present express train service, we realise how vast is the progress we have made.

Horace Walpole was travelling in Kent in August, 1752, and he gives a most graphic picture in his *Letters* of the dangerous state of the Kentish roads. On August 3 of that year he writes to a friend: "We have had piteous distresses" in the journey through Kent. When he arrived at Penshurst, he exclaims: "Now begins our chapter of woes....The only man in the town who had two horses would not let us have them, because the roads, as he said, were so bad. We were forced to send to the Wells for others, which did not arrive till half the day was spent."

The journey was made from Penshurst to Lamberhurst, and then Walpole writes: " Here our woes increase. The roads grew bad beyond all badness, our guide frightened beyond all frightfulness. However, without being at all killed, we got up, or down, I forget which, it

At Penshurst

was so dark, a famous precipice called Silver Hill and about ten at night arrived at a wretched village called Rotherbridge. We had still six miles hither, but determined to stop, as it would be a pity to break our necks before we had seen all we intended. But, alas! there was only one bed to be had ; all the rest were inhabited by smugglers, whom the people of the house called mountebanks. We did not at all take to this society, but armed with links and lanthorns, set out again upon this impracticable journey. At two o'clock in the morning we got hither to a still worse inn, and that crammed with excise officers, one of whom had just shot a smuggler."

The modern improvement in road-making in Kent dates from about 1802. In that year Mr Rennie, the famous engineer, was surveying the Weald with a view to the construction of good roads. He found the country almost destitute of suitable roads, and the interior was practically untraversed except by bands of smugglers, who kept the country-side in a state of terror.

Since that date new roads have been constructed and the old main roads improved. The principal highways that traverse Kent start from London, and stretch across the country in the direction of east and south-east, thus connecting the metropolis with the coast towns. The most important is the Dover Road, which crosses Blackheath and follows the route of Watling Street to Dover. Another trunk-line leaves the Dover Road at New Cross and thence proceeds through Maidstone and Ashford to Hythe. The Hastings Road leaves London and proceeds by way of Bromley, Sevenoaks, and Tonbridge to Hastings. Two short but important roads diverge from Canterbury, one going to Ramsgate and Margate, and the other to Sandwich and Deal. Besides these main roads, there are numerous cross-roads that connect these with other high roads.

The nineteenth century saw the introduction of railways, and one of the first lines to be opened, in 1838, was that from London to Greenwich. Kent is now served by the South-Eastern and Chatham Railway, which is an amalgamation of the South-Eastern and the London, Chatham, and Dover Railways. The former line was opened in 1844 and the latter in 1859, while the amalgamation took place in 1899.

Knole House, Sevenoaks

The canals of Kent are quite unimportant, and only two are worth a short notice. The Royal Military Canal runs for 23 miles from the Rother to Hythe. It was constructed during our wars with Napoleon, but is now in a neglected condition. The Gravesend and Rochester

Canal, now the property of the South-Eastern and Chatham Railway, was cut in the early years of the nineteenth century to shorten the river-journey from Strood. It was never very successful and now ends at Higham, the bed of the canal from that point forming the railway track.

28. Administration and Divisions — Ancient and Modern.

We shall understand the government of the county of Kent all the better if we first learn something about its earliest mode of administration. Of course, many alterations have been made since the time of our Saxon forefathers, who introduced a new form of government when they conquered Britain. But we find all through our history that when some alterations have been made in the mode of county government, there has always been great care to graft new ideas on to the old institutions, rather than to uproot or supplant the latter.

Then, as now, the government of each county was partly central and partly local. The chief court of the county of Kent in the earliest times was the Shire-mote, which met twice a year. It had two chief officers, known as the Ealdorman and the Sheriff, the last of whom was appointed by the King.

Then our Saxon forefathers divided Kent into Hundreds, each of which probably contained at first one hundred free families. There are 73 hundreds in Kent, which are grouped into five lathes (from an Anglo-Saxon word meaning a division of land), known as Sutton-at-Hone, Aylesford, Scray, St Augustine, and Shepway. Each hundred had its own court, the hundred-mote, which met every month for business. Each hundred was divided into townships, or parishes, as we now call them. Each township had its own assembly, or *gemot*, as it was then called, where every freeman could appear. This assembly made laws for the township, and it also appointed officers to enforce these by-laws, or laws of the town. These officers were the reeve, and the tithing-man, who was a constable, something like our policeman. The court of the township was held whenever necessary, and the reeve was the president, or chairman.

Now let us come down to our present mode of county government. The chief officers in the county are the Lord-Lieutenant and the High Sheriff. The former is generally a nobleman or rich landowner, who is appointed by the Crown; while the latter is chosen every year on "the morrow of St Martin's Day," November 12th. The County Council now conducts the main business of the county, and since 1888 the Council House has been at Maidstone, the county town. This County Council consists of 24 Aldermen and 72 Councillors. The latter are elected to their office, while the former are co-opted. The County Council has most important business to transact. It keeps the high roads and bridges in good repair; it appoints the police; it manages lunatic asylums and reformatories; and generally carries into effect the laws passed by Parliament.

The County Council represents the central form of county government, which was started in 1888 ; but for local government in the towns and parishes another Act was passed in 1894, and new names were given to the local bodies.   In the large parishes the chief authorities

Rochester Castle

are now called the District Councils, of which there are 22 in Kent, while the smaller parishes have their Parish Councils, or parish meetings.   There are, however, some towns in Kent that are called boroughs, and these have larger and different powers of government than the parishes.   Canterbury is called a county borough, and has the power of a county council.   The oldest boroughs in Kent are Deal, Dover, Faversham, Folkestone, Gravesend, Hythe, Lydd, Maidstone, Margate, New Romney, Queenborough, Rochester, Sandwich, and Tenterden. The more recent boroughs are Chatham, Gillingham, Ramsgate, and Tunbridge Wells.

Kent is also divided into 26 Poor Law Unions, each of which is under a Board of Guardians, whose duty it is to manage the workhouses and appoint various officers to carry on the work of relieving the poor and aged.

For purposes of justice Kent has two quarter sessions, which meet at Canterbury for East Kent, and at Maidstone for West Kent ; and a number of petty sessions, each having magistrates or justices of the peace, whose duty it is to try cases and punish offenders against the law.   In Kent there are 419 civil parishes, so that it will be seen that there must be a large number of magistrates.

For ecclesiastical purposes Kent has two sees, Canterbury and Rochester, the former being the seat of the Archbishop and the latter of the Bishop.   Each see is divided into archdeaconries, rural deaneries, and parishes. An ecclesiastical parish is not always the same as a civil parish, and there are more of the former than the latter.

For purposes relating to education, there are 17 Education Committees for the larger towns and boroughs. An education committee is appointed by the Kent County Council, and this has the management of the schools in the rest of the county, both elementary and secondary.

Kent is represented in the House of Commons by 15 members of parliament. The following seven parliamentary boroughs each send one member :—Canterbury, Chatham, Dover, Gravesend, Hythe, Maidstone, and Rochester. Kent is divided into eight divisions for parliamentary representation, and each of these sends one member.

Canterbury Cathedral from the Meadows

29. The Roll of Honour of the County.

It would be difficult to find another county that has such a long and distinguished roll of honour as the "Royal County" of Kent. Let us glance at a few of the Kentish worthies, some of whom were born in Kent, while others lived and died in that county.

Kent has been well named the royal county, for many of our kings and queens have been natives of that county. Henry VIII was born at Greenwich Palace, as were also his two daughters, who afterwards became Queen Mary and Queen Elizabeth. Henry VIII often visited Hever Castle, where Anne Boleyn lived with her father. James I's two daughters, the Princesses Mary and Sophia, were both born at Greenwich Palace; and it was King William III who converted this palace into a hospital for seamen.

When we think of Canterbury and Rochester we naturally connect the two cathedrals with the names of a long line of archbishops, bishops, and deans. From this list the name of Becket stands out, for Canterbury Cathedral was the scene of his martyrdom on that dark December evening in 1170.

Among great English statesmen Kent has been the home of Walsingham and Sir Nicholas Bacon, who were two of Queen Elizabeth's chief advisers. Lord Chatham, the great statesman and orator, lived at Hayes Place, near Bromley, and there, too, his son, William Pitt, was born in 1759. The younger Pitt was further connected with Kent, for he became Lord Warden of the Cinque Ports, and often resided at Walmer Castle. Many of our statesmen since Pitt have been Lord Warden, but the Duke of Wellington, our greatest general, was the most famous. For many years he spent much of his time at Walmer Castle, where he died in 1852.

James Wolfe, who gave us Canada, was born at Westerham in 1727. He gained his victory at the cost of his life, and his body was brought home and buried at Greenwich Church. General Gordon was also connected with Kent, for he was born at Woolwich, and his good

Fort House, Gravesend
(General Gordon's Home)

work is still remembered at Gravesend, where a monument has been erected to this great hero.

Kent has on its roll of honour a number of distin-

Sir Philip Sidney

guished historians and antiquaries. William Camden, the writer of *Britannia*, or an account of the British Isles, lived at Chislehurst, where he died in 1623. Lambarde, the author of *The Perambulation of Kent*, and Hasted, who wrote the best *History of Kent*, were both residents in this county. Grote, the historian of Greece, was born at Beckenham, and was educated at the grammar school at Sevenoaks. Henry Hallam, the author of the *Constitutional History of England* and other historical works, lived in Kent, and died at Pickhurst, near Hayes, in 1859.

Part of Garden front, Penshurst Place

When we come to poets who have made their home in Kent, or who have been associated with it, we find the names of Gower and Chaucer, Sidney and Wyatt, Marlowe and Lovelace, while our present poet-laureate,

Alfred Austin, lives at Swinford Old Manor, near Ashford.   Chaucer has made Kent famous by his *Canterbury Tales*, which describe the personages who went from the Tabard Inn at Southwark to Becket's Shrine.   Wyatt lived at Allington Castle, while the name of Sidney will ever be associated with Penshurst, where he was born in 1554.   Marlowe, a contemporary of Shakespeare, was born and educated at Canterbury.   There is a memorial to this dramatist in the centre of the Old Butter Market in his native town.

Lovelace, the cavalier and poet, was imprisoned for supporting the "Kentish Petition."   It was then that he wrote the well-known stanzas, beginning

> Stone walls do not a prison make,
> Nor iron bars a cage;
> Minds innocent and quiet take
> That for an hermitage.

The family of Lovelace had property at Canterbury as well as at Bethersden.

William Caxton, the first English printer, was a native of Kent.   In 1422, he says, " I was born and learned my English in Kent, in the Weald, where English is spoken broad and rude."   The introduction of printing into England by Caxton was one of the greatest events in the history of our nation.

Among other literary men we can mention only Hooker, Evelyn, Barham, and Dickens.   The "judicious" Hooker, a learned divine of the Elizabethan period, was rector of Bishopsbourne, and is now remembered by his learned work, *The Laws of Ecclesiastical Polity*.   John

Evelyn, the author of the celebrated *Diary*, lived at
Deptford.   It was at Evelyn's place at Sayes Court that
Peter the Great resided when learning the art of ship-
building.   Richard Barham, born at Canterbury in 1788,
was the author of the *Ingoldsby Legends*, a series of
humorous metrical tales.   Dickens, the great novelist of

Chilham Church

the Victorian era, lived in his early days at Chatham,
and knew well the district around Rochester.   He spent
the later years of his life at Gad's Hill Place.

We must let the name of Sidney Cooper stand alone,
for he is essentially a Kentish artist.   He was born at
Harbledown, near Canterbury, and excelled as a painter
of cattle.   A school of art was founded by him at
Canterbury, and forms a fitting memorial to his work.

There now remain to be mentioned two of the greatest men of science. William Harvey, the discoverer of the circulation of the blood, was born at Folkestone in 1578, and in that town there is a fine statue to his memory. Charles Darwin, whose book *The Origin of Species* first clearly sets forth the principle of natural selection, lived at Down in the neighbourhood of Orpington. He may be said to have wrought a greater change in scientific thought than anyone since the day of Newton.

Chiddingstone

30. THE CHIEF TOWNS AND VILLAGES OF KENT.

(The figures in brackets after each name give the population in 1901, and those at the end of the sections give the references to the text.)

Ashford (12,808) is a market-town and the centre of an agricultural district. It is important as the junction of four lines of railway, and as having the locomotive and carriage works of the South-Eastern and Chatham Railway. It is mainly of modern growth; but the fine Parish Church is noticeable for its tower, and the College and Grammar School date from the 17th century. Shakespeare calls Jack Cade the "headstrong Kentish man, John Cade of Ashford." (pp. 17, 69, 90, 94, 101, 128.)

Aylesford (2678) stands on the Medway, which is spanned by a fine stone bridge. Kit's Coty House is in the neighbourhood, and there are large blocks of stone lying between it and the town which are probably the remains of circles of stones. (p. 95.)

Beckenham (26,331) is now a populous suburb of London, with a modern church and many large houses. (p. 127.)

Bexley (12,918) stands on the river Cray, in the midst of beautiful country. A church has existed on the site of the present building since the 8th century. In the vicinity are Deneholes.

Birchington (2128) is a quiet and fashionable sea-side resort between Herne Bay and Westgate. Dante Gabriel Rossetti, poet and painter, died here in 1882. (pp. 36, 79.)

Broadstairs (6466) is a well-known watering-place nearly two miles north-east of Ramsgate. Queen Victoria often resided here in her early days, and Charles Dickens lived at Bleak House.

9—2

Bromley (27,354) is a market-town on the river Ravens-bourne. It stands on high ground in the midst of a richly wooded and picturesque country. The Bishops of Rochester used to live at Bromley Palace; and near by stood the stately buildings of Bromley College. (pp. 16, 124.)

Aylesford Bridge and Church

Canterbury (24,899), on the Stour, is the ecclesiastical capital of England. The Archbishop is the Primate of All England and crowns the new Sovereign in Westminster Abbey. Besides its importance as a cathedral city, it is a municipal and parliamentary borough, and a county by itself. It is probably the oldest seat of

Christianity in England, and the ancient Church of St Martin, built partly of Roman brick, has been called the Mother Church of England. There are some remains of the ancient city walls; but of six gates, West Gate is the only one now standing. The Norman Castle has been converted into gas-works; and the old Guildhall has been rebuilt. The Cathedral stands on the site of an early church, perhaps built by the Britons. The first church was rebuilt by Lanfranc, and much enlarged by Anselm, under the care of Prior Ernulph. Conrad finished the chancel, and it was in this glorious building that Becket was murdered. This cathedral was destroyed by fire, and rebuilt by " English " William in 1184. The nave and transept were rebuilt in 1378, and the grand central tower was added in 1495. The present cathedral is much as it was 400 years ago. Canterbury is not a manufacturing or a commercial centre, but it has a large trade in grain and hops and manufactures of linen and worsted. The King's School was founded by Henry VIII, and has educated Marlowe, Harvey, Lord Tenterden, and other great men. St Augustine's College, a beautiful building, stands on the site of an ancient monastery of which the noble entrance gate built in 1300 alone remains. Many events of great interest are associated with Canterbury, the very name of which calls up memories of Chaucer's *Canterbury Tales*. Nowadays, the city is famous for its cavalry and infantry barracks, and every year it has its " Cricket Week." (pp. 3, 7, 17, 58, 67, 71, 88, 94, 101, 102, 122, 123, 124, 128, 129, 130.)

Chatham (38,504) stands on the Medway and is both a municipal and parliamentary borough. The chief features of the town are the Dockyard, Barracks, Convict Prison, and Hospitals. The Dockyard is one of the most famous in the kingdom. Chatham Lines are elaborate fortifications, consisting of trenches, batteries, and subterranean passages. For purposes of defence there is a chain of modern forts. Chatham is a town of some antiquity and Roman remains have been found in its neighbourhood. (pp. 16, 60, 67, 68, 79, 122, 129.)

Chislehurst (7429) is eleven miles from London and a favourite residence of London merchants. It has one of the most beautiful commons, surrounded by beautiful trees, and about 300 feet above the sea.   Camden, Walsingham, and Bacon lived here; and Napoleon III died in exile at Camden House.   (pp. 101, 126.)

Cranbrook (3949) is the principal market town of the Weald. It was once the centre of the clothing trade, introduced by the Flemings in the reign of Edward III.   The work ceased early in the 18th century, but there are still remains of the picturesque cloth-halls.   (pp. 58, 71, 74, 101, 112.)

Crays, The, are four parishes situated on the little river Cray, above Bexley.   The scenery of the Crays is varied and pleasing. There are fine woods near by, and hop-gardens, fruit-farms, and paper-mills are numerous.   The order of the Crays in descending the river is as follows: **St Mary Cray** (1894), **St Paul's Cray** (1207), **Foot's Cray** (5817) and **North Cray** (661). (p. 69.)

Darenth (3493) stands on the river Darent, two miles south-east of Dartford.   In the neighbourhood, some extensive Roman buildings have been discovered, and an Anglo-Saxon burying-place.

Dartford (18,644) on the river Darent, has important powder-mills and paper-mills.   The first paper-mill is said to have been erected here by Sir John Spielman.   Wat Tyler's insurrection of 1377 began at Dartford.   (pp. 16, 26, 69, 88.)

Deal (10,581) is a borough, market-town, seaport, and pleasant watering-place on the south-east coast.   It is also famous as a pilot-station, and is opposite the Goodwin Sands.   It is thought by many that Julius Caesar landed at Deal in 55 B.C.   The Castle was built by Henry VIII for the defence of the coast.   (pp. 38, 76, 79, 80, 107, 122.)

Deptford (110,122) is three miles from London Bridge, and stands on the river Thames and river Ravensbourne.   It was once a Royal Dockyard, but is now the City cattle market.   John Evelyn, the diarist, lived at Sayes Court; and it was at Deptford that Peter the Great learnt the art of shipbuilding.   (pp. 6, 16, 31, 129.)

Dover (41,794) is one of the most famous of the Cinque Ports and the terminus of the South-Eastern and Chatham Railway. The town is ancient and the fine Castle recalls stirring memories of the past. The Pharos, or Watch-tower, is a Roman building, and the Church of St Mary is of great antiquity. Dover has a good harbour and extensive works are now in progress to increase its importance. It has considerable trade with the Continent, and there is a good steam service to Ostend and Calais. The cliffs are famous and on the heights are extensive fortifications and barracks. In the neighbourhood, borings have been made for the proposed Channel Tunnel, and also in search of coal. (pp. 11, 38, 42, 50, 58, 73, 75, 79, 80, 83, 94, 100, 102, 106, 114, 115, 122, 123.)

Eltham (7226) is a rapidly growing suburb of London, about three miles south-east of Blackheath. In the ruins of Eltham Palace, which was the residence of our Kings from Henry III to Henry VIII, is a very fine banqueting hall. (p. 108.)

Erith (25,296) stands on the Thames, about four miles east of Woolwich. There are ruins of an abbey, which was once famous as Lesnes Abbey. Now the place has a gun factory, gunpowder factory, iron-works, and engineering works. (pp. 33, 69.)

Faversham (11,625) is a borough, market-town, and port on the river Swale. The principal industry is the oyster fishery, but there is some trade in timber, coal, and hops, and there are manufactures of bricks and cement. (pp. 67, 69, 83, 96, 122.)

Folkestone (32,150) is a borough, market-town, seaport, and fashionable watering place on the south-east coast. It is the port for the steam service to Boulogne, and has considerable trade with the Continent. The Leas is a grassy expanse on the top of the cliffs; and the Warren is the picturesque undercliff to the east of the old town, which still has considerable interest in the coast fisheries. The first nunnery in England was built here by St Eanswith; and the large Parish Church is a very beautiful building, with fine wall decorations. Harvey was born at Folkestone and there is a statue to his memory on the Leas. (pp. 11, 23, 39, 40, 73, 75, 80, 85, 122, 130.)

Gillingham (41,441) is a borough of modern growth. It stands on the Medway, north-east of Chatham, and is celebrated for its cherry-orchards. (pp. 60, 62, 67, 122.)

Gravesend (11,662) is a borough, market-town, and river-port, about 24 miles from London by rail. It is the port of London and a pilot station. It has facilities for boating and yachting, but its chief industry is fishing. (pp. 33, 79, 122, 123, 126.)

Greenwich (67,315) is a borough on the Thames about 3½ miles from London Bridge. Greenwich was a royal residence as early as 1300, and several of our monarchs were born in the palace or lived there. Greenwich Hospital was built on the site of the palace and was long a home for disabled seamen. This splendid building is now the Royal Naval College. Greenwich Observatory, built on high ground in the Park, is world-famous. From it we reckon our longitude; and its astronomers make most careful astronomical, magnetic, and meteorological observations. Greenwich has several manufactures, and telegraph and engineering works. (pp. 6, 11, 32, 56, 78, 88, 123, 124.)

Herne Bay (6726) is a seaside resort, occupying an agreeable position on the north coast. (pp. 25, 36, 40, 41, 46, 56, 94.)

Hever (723) stands on the Medway, seven miles west of Tonbridge. The most important building is Hever Castle, which was built in the reign of Henry VI and was recently restored. Henry VIII used to visit Anne Boleyn at this castle, and her father is buried in the church. (p. 101.)

Hythe (5557) is one of the Cinque Ports, but there is now a stretch of shingle, a mile wide, between the town and the sea. It is well known from the Government School of Musketry and its rifle butts. (pp. 22, 39, 40, 47, 83, 94, 101, 106, 118, 122, 123.)

Lee (127,495) is at the southern end of Blackheath. What was once a pleasant village has been built over and now forms part of London.

Leeds (650) is five miles north-east of Maidstone. The Castle, a stately building, stands in the centre of a finely wooded park and is the great attraction of this part of Kent.   The main portion dates from the 13th century, but much of the present building is modern. (p. 104.)

Lewisham (128,346) lies to the south of Greenwich and is really part of London.   (pp. 6, 25.)

Lydd (2675) is a borough and market-town a little to the south-west of New Romney.   Lydd is an ancient town and a member of the Cinque Port of Romney.   In the vicinity is a military camp.   The "lyddite" shell is manufactured here—hence its name.   (pp. 42, 69, 122.)

Lyminge (1030) is to the north of Hythe, and has one of the most ancient churches in Kent.   (pp. 17, 100.)

Lymne (467), west of Hythe, was famous in Roman times. The Romans had an important station here, and the ruins of their fortress are still to be seen.   (pp. 40, 44, 96, 114.)

Maidstone (33,516), on the Medway, is the county town of Kent.   The chief objects of interest are the very large and important church, the ancient palace of the archbishops, and the college.   In mediaeval and later history it has been associated with the revolts of Wat Tyler, Jack Cade, and Sir Thomas Wyatt.   In 1648 Fairfax stormed and captured the town after a stout resistance. There are industries connected with paper-making, brewing, malting, iron-founding and the making of agricultural implements.   The chief trade, however, is in hops, Maidstone being the centre of the hop district.   (pp. 16, 23, 67, 69, 73, 90, 91, 120, 122, 123.)

Margate (23,118), is a borough and popular seaside resort in the Isle of Thanet.   Bathing machines were invented and first used here about 1750.   Owing to its excellent sands and good air, it is thronged in the season with visitors chiefly from London. (pp. 36, 46, 53, 56, 76, 122.)

Milton (next Sittingbourne) (7086) has a fine church that succeeded a Saxon building.   (p. 69.)

Minster (2338) in Thanet, has an ancient church that was erected by the monks of St Augustine.

Minster (1306) in Sheppey, has an ancient church of great interest. Remains of an abbey, built in the seventh century, still stand near the church. (pp. 41, 96.)

New Romney (1328) was once a famous Cinque Port, but is now without a harbour. It has little trade, but is celebrated for its great sheep fair in August. (pp. 18, 39, 47, 122.)

Northfleet (12,906), on the river Thames, is in the parliamentary borough of Gravesend. There are large Portland cement works, lime and brick works in the neighbourhood. (pp. 33, 69.)

Orpington (4259) is three miles south-east of Chislehurst. Much fruit is grown for the London markets and hops are also cultivated. (p. 130.)

Penshurst (1678) stands at the junction of the Eden with the Medway. Penshurst Place is one of the famous mansions of Kent, and is associated with the Sidneys. Penshurst village is very picturesque. (pp. 16, 110, 112, 115, 128.)

Queenborough (1544) is a borough on the river Swale in the Isle of Sheppey. It once returned two members to Parliament. There is an oyster fishery ; and a daily service of steamers runs to Flushing. (pp. 76, 85, 122.)

Ramsgate (16,503) is a popular watering-place in Thanet. There is a good harbour and the fine pier was built by Smeaton in 1750. Some ship-building is carried on and the fishery is important. (pp. 36, 75, 79, 95, 122.)

Rochester (14,520) is a city, borough, and port on the Medway. It was of importance in British, Roman, and Saxon times, and its see was founded by Ethelbert. The cathedral was commenced about 604 A.D. and rebuilt in the 11th, 12th, and 13th centuries. The castle keep still stands and is one of the finest in our land. Rochester has industries connected with machinery, and many of the people are employed at Chatham Dockyard. There is also some shipping and fishing. The city is connected with Strood by a fine bridge over the Medway. (pp. 3, 16, 67, 106, 122, 123, 124.)

Sandgate (2023) is a quiet little watering-place to the south-west of Folkestone.   The coast is protected by a sea-wall. Sandgate Castle was built by Henry VIII for the defence of the coast.   (pp. 39, 48, 107.)

Sandwich (3170) is a borough and Cinque Port on the river Stour.   It was a seaport of great importance till the reign of Edward VI, when its harbour was choked with sand.   The buildings in the town are quaint and picturesque, and tell of its past history.   (pp. 38, 40, 58, 83, 101, 122.)

Sevenoaks (8106) is a market-town, pleasantly situated on high ground in the midst of fine scenery.   Knole House is one of the most interesting places in Kent.   The Grammar School dates from the 15th century.   (pp. 108, 127.)

Sheerness (18,179) is a seaport, dockyard, and naval arsenal on the north-west of the Isle of Sheppey.   It is now strongly fortified and the harbour is safe and commodious.   The Dutch captured Sheerness in 1667 ; and the Mutiny of the Nore broke out here in 1797.   (pp. 31, 34, 67, 68, 76, 79.)

Sittingbourne (8943) is a seaport and market-town of great antiquity on Milton Creek, a branch of the Swale.   Its trade is chiefly in bricks, corn, and coal.   (p. 69.)

Strood (10,006) on the left bank of the Medway, is connected with Rochester by a fine bridge.

Tenterden (3243) is an ancient borough and market-town to the north-west of Appledore.   Its church was built from funds originally devoted to the maintenance of the sea-walls on the Goodwin Sands.   (p. 122.)

Tonbridge (14,054) stands on ground rising from the Med-way.   The ruins of the castle are near the centre of the town.   The Grammar School is an ancient foundation dating from the 16th century.   (pp. 16, 22, 69, 104, 110.)

Tunbridge Wells (31,549) is a borough, market-town, and popular inland watering-place, situated in the midst of picturesque scenery. The chief parade is the Pantiles, and at the end are the chalybeate springs, discovered by Lord North in 1606. Tunbridge Wells has long been a fashionable resort, and has associations with Queen Henrietta, Dr Johnson, Beau Nash, and other celebrities. (pp. 22, 53, 56, 122.)

Walmer (5248) is a pleasant little town and a favourite summer resort. It has long been famous for its castle, which was built by Henry VIII. Until lately, it was the official residence of the Lord Warden of the Cinque Ports. (pp. 84, 107, 124.)

Eastry

Westerham (2905) is a market-town to the west of Sevenoaks. General Wolfe was a native, and there is a monument to him in the Church. (pp. 11, 124.)

Westgate-on-Sea (2738) is a pleasant watering-place of modern growth in the Isle of Thanet. (pp. 36, 46.)

Whitstable (7086) is the port of Canterbury.   It has considerable trade, and its oyster fisheries have been famous since Roman times.   (pp. 34, 40, 78.)

Woolwich (117,178) is a borough on the Thames.   The chief feature of Woolwich is the Arsenal, one of the most extensive and complete in the world.   Among the other buildings of interest are the Military Academy, and the Herbert Hospital.   (pp. 6, 9, 25, 32, 68, 124.)

Wye (1312), to the north-east of Ashford, has a large and handsome Church.   The South-Eastern Agricultural College is at Wye.

Fig. 1. The area of Kent compared with that of
England and Wales

Fig. 2. The population of Kent compared with that
of England and Wales

Density of population   Density of population of   Density of population
  of Lancashire   England and Wales   of Kent
(1070 to sq. mile)   (558 to sq. mile)   (618 to sq. mile)

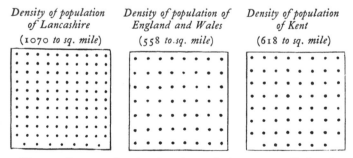

Fig. 3. Comparative density of population to sq. mile (1901)
(Note.   Each dot represents 10 people)

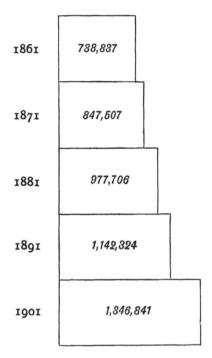

Fig. 4. The growth of population in the Ancient County
of Kent, from 1861—1901

(Note. Part of the Ancient County of Kent is now comprised in the
County of London)

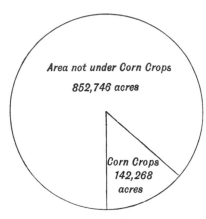

Fig. 5. This diagram shows the proportionate area of Kent growing Corn Crops—(Wheat, Barley, Oats, Rye, Beans and Peas) (1905)

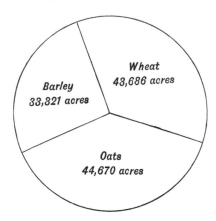

Fig. 6. This diagram shows proportionate cultivation in acres of Oats, Wheat, and Barley (1905)

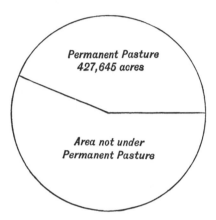

Fig. 7. This diagram shows the proportionate area of Permanent Pasture to area of the County of Kent (1901)

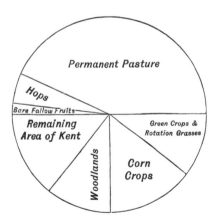

Fig. 8. This diagram shows the distribution of Crops, Permanent Pasture, etc. in Kent (1905)

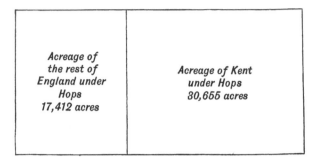

Fig. 9. This diagram shows the comparative areas of
England and Kent under Hops (1905)

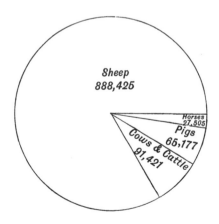

Fig. 10. This diagram is a comparison of the numbers of
Sheep, Cows and Cattle. Pigs, and Horses in Kent